HELICOPTERS AT WAR

HELICOPTERS AT WAR

A Pictorial History

by

DAVID W. WRAGG

ROBERT HALE · LONDON

© David W. Wragg 1983
First published in Great Britain 1983

ISBN 0 7090 0858 9

Robert Hale Limited
Clerkenwell House
Clerkenwell Green
London EC1R 0HT

Photoset and printed in Great Britain by
Redwood Burn Ltd, Trowbridge, Wiltshire
Bound by WBC Bookbinders Ltd.

Contents

Acknowledgements

I should like to thank those who assisted in the research and
preparation of material used in this book, and in particular those who
assisted in the collection of the illustrations, including:
Novosti
Fleet Air Arm Museum
Royal Air Force Museum
Science Museum
Central Office of Information
Ministry of Defence
Agusta
Bell
Boeing-Vertol
MBB
PZL
Hughes
Lockheed
Sikorsky
Westland

1
TRIAL AND ERROR

There are some inventions which leave the proud inventor to tell an incredulous world that an application has still to emerge, or even that he did it merely in the interests of science, but obviously the most useful invention is that which fulfils a long-felt need. This may be too practical a statement, too unimaginative or even myopic, but in the event, it is at least pleasant to record that the helicopter came firmly into the latter category of invention. This goes some way towards explaining why the truly practical helicopter did not hang around in cold storage waiting for a suitable application or an interested customer, but instead enjoyed an immediate appreciation of its practical applications which stimulated further trials to ascertain its value for other relatively marginal roles.

It is necessary to take a look to see just why this was so, and why it took so long for the helicopter to emerge as a machine with a worthwhile role some forty years after the Wrights brought conventional aeroplane flight to reality. Even today, a further forty years or so later, the helicopter is still amongst the most difficult machines to fly, the most expensive to buy, maintain and operate, and yet remains amongst the slowest and with the poorest lift and endurance for the power available. Indeed, the problems over power and the quality of structure required for success to be possible, did little to hasten the advent of the helicopter, although the real reason lay in the inability of the early pioneers to ascertain the control requirements and characteristics of the machine.

The value of the helicopter, today as always, lies in its versatility, the ability to take off and land vertically, to hover in mid-air and even to move backwards or sideways if necessary. Surprisingly, in the early days of experiment, the helicopter, and other machines vaguely like it, were not seen in this light. The helicopter was thought of as a short cut towards powered flight and often even confused with the alleged benefits of the ornithopter, which was supposed to achieve flight by emulating the birds and flapping its wings. This simply underlines the difficulty which existed in finding a realistic appreciation of the problems of control inherent in the helicopter, which went hand in hand with little or no realization of its peculiar benefits. The early pioneers sometimes viewed some kind of movement in the wings as a substitute for adequate momentum during the take-off or run, and even as an alternative to a prime mover with an adequate power-to-weight ratio, while rotation of the wings was merely a simpler substitute for flapping!

The concept of the glider and the aerostat, or balloon, predated that of the helicopter, but even these more mundane projects were still far from reality when the Italian artist, Leonardo da Vinci, produced what has since come to be regarded as the first helicopter design. Leonardo's helicopter was to consist of a helical screw which would have been powered by a clockwork motor. The design appeared during the middle years of the artist's life, around 1500. Leonardo's other designs for flying machines were intended for man power, with the ornithopter concept being strongly favoured.

A model of Leonardo's helicopter design, which first appeared around 1500.

It was not until 1784, a year after the first balloon ascents, that a more practical helicopter design first appeared, with two Frenchmen, Launoy and Bienvenue, producing a small helicopter with two rotors at either end of a shaft, and the whole contraption, although it was indeed rather simple, powered by a bowstring! The limitations were obvious, but it did at least provide the basis for a successful toy.

While the helicopter was so long in appearing, applications for the machine came rather faster. The arrival of both the hot-air balloon and the hydrogen balloon, within weeks of one another during late 1783, soon led to the captive observation balloon being used by the military; captive only so that the observation balloon could be prevented from drifting over enemy-held territory, with the attendant risk of a forced

Aerial observation, by balloon! The Battle of Fleurus, 1794, in an illustration from a snuff-box lid.

landing! The first recorded use of balloons for artillery spotting from the air came in 1794 at the Battle of Maubeuge, and was followed later that same year by the use of observation balloons at the Siege of Charleroi and the Battle of Fleurus. Here was the beginning of one of the more important roles for the helicopter on the battlefield, that of the aerial observation post or AOP. This role was to be the preserve of light aircraft while the helicopter was awaited, and for many years afterwards as the early helicopters only gradually overtook light aircraft in their development; this was to be particularly true for the

Airborne assault, by balloon! An artist's flight of fancy for a French invasion of Britain.

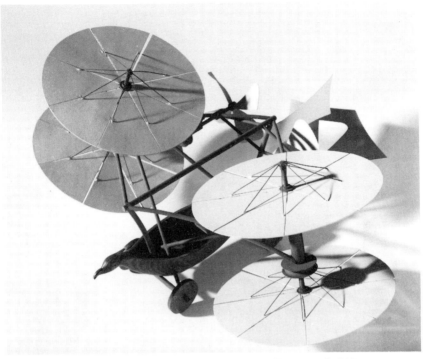

Sir George Cayley's "aerial carriage" compound helicopter of 1843.

armies of the poorer nations, which had to struggle to afford aviation's
new arrival in useful quantities.

The famous British pioneer, Sir George Cayley, designed a
helicopter model in 1796, rather on the lines of that produced by
Launoy and Bienvenue, but one cannot be sure just how aware he was
of their work, given the poor communications of the period. Cayley at
this stage was just twenty-three years old, and may have heard reports
of the work of the two Frenchmen, which he was attempting to assess
for himself. Cayley, a genius with interests in many fields including
politics and social reform, seems to have ignored the helicopter for
some years after this (although he did not abandon gliders and airships
in the intervening period), before publishing a design in 1843 for an
"aerial carriage", with coaxial rotors or rotating wings for lift and a
pusher propeller for forward flight. This was never built, and there are
some who even maintain that the design was based on the work of
another Englishman, Robert Taylor, but Cayley was generally far
ahead of his contemporaries by this time and had already identified

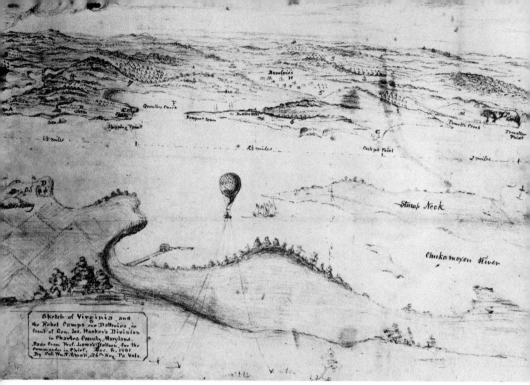

A sketch of Virginia during the American Civil War, made from a balloon in December 1861.

many of the problems affecting heavier-than-air powered flight, as well as some of the solutions. Later, he built and flew first a boy-carrying and then a man-carrying glider kite.

There were other helicopter designs during this interesting, but for all practical purposes fruitless, period of aeronautical history. Cassius, in 1845, produced a triple-rotor design with a large central rotor for lift and two smaller rotors for steering control; it was never built. Another Englishman, W. H. Phillips, built a steam-powered model helicopter with rotor-tip steam jets using a fuel consisting of a mixture of charcoal, nitre and gypsum, and this actually managed to fly across two fields, although their size is not known! His fellow countryman, Bourne, built and tested a number of clockwork-powered helicopter models based on the double-rotor Launoy and Bienvenue concept.

A Frenchman, the Vicomte Ponton d'Amecourt, built a number of helicopter models, some powered by steam engines while others used clockwork motors, and although his first, in 1861, was a failure, many of those which followed, with their contra-rotating rotors, flew extremely well. One even had a parachute, so that it could return to earth safely after the motor wound down, while the contra-rotating

12

As an alternative to the balloon for observation duties, the British Army
experimented with man-carrying kites. The American pioneer Samuel Cody
is on the horse.

rotors solved the problem of the fuselage or main body of the
helicopter spinning uselessly and fatally on its own rotor. More
ambitious, had it been built, was the design for a large steam-powered
helicopter which may even have influenced the famous author, Jules
Verne, in writing his *Clipper of the Clouds*, published in 1886. De la
Landelle made his own contribution to the language by coining the
words "aviation" and "aviator" – "*l'aviation*" and "*l'aviateur*" – in
1863.

Rather more practical was the Paucton Pterophore of 1868, another
French design, but for a machine with two large rotors or
"pterophores", one for lift and the other for propulsion. Again,
however, this was a design which never left the drawing-board, but
this was probably as well, since the fatal weakness would have been the
use of a man-powered crank for propulsion.

Another rather over-rotored design was produced in 1874 by
Achenbach, featuring a large rotor for lift, a small rotor for countering
the rotational effects of the main rotor on the fuselage, and small rotors
for steering and propulsion. This was never built.

A pedal-powered helicopter built in 1872 by Renoir, failed to fly.

Three more important nineteenth-century helicopter designs, one with a contra-rotating rotor by Dieuaide in 1877, a coaxial design, that is with two rotors on the same axis, by Castel, and a large contra-rotating design by Forlanini, both in 1878, remained far short of the practical helicopter. Emmanuel Dieuaide's design used large contra-rotating rotors powered by steam piped from boilers on the ground, and was a novel if extremely limited means of overcoming the weight of a boiler, which was just one of the disadvantages of steam-power in flight. Just as novel, and limited, was Forlanini's method, which consisted of pre-heating a boiler on the ground and attaching it to the helicopter, leaving the steam to flow through to the pistons, but this worked sufficiently well for a model to rise to the dizzy height of forty-two feet on one occasion!

Gustav Trouve built a model ornithopter in 1870, powered by twelve blank revolver cartridges fired automatically. This model succeeded in flying for 200 feet after a mid-air launch from a balloon. No doubt the flight was accompanied by spectacular, if not alarming, sound effects, and perhaps it was just as well that Trouve abandoned plans for a man-carrying version. Although this was probably the first ornithopter to fly, the claim to flight did not meet the strict scientific criteria for flight, since the machine did not take off under its own power and was not under control during flight. Later, in 1887, Trouve built an electrically powered ornithopter using a lightweight electric motor, which received its current through a pair of wires trailing on the ground. Inspired by this success, Trouve planned a larger version, which was probably never built and almost certainly never flew, but for which he foresaw a "value for military use such as observation and reconnaissance".

The idea of returning models safely to the ground with the aid of a parachute appeared again in 1877 when Melikoff built a helicopter with a six-bladed rotor and what he described as a "gas-turbine" engine, in fact more akin to the concept of a Wankel rotary engine. A contemporary, Dandrieux, built rubber-band-powered models in 1878 and 1879.

Powered heavier-than-air flight finally arrived in December 1903, with the successful flights by the two Wright brothers, using the aircraft which later became known as the Flyer 1 for a series of short flights at the Kill Devil Sands, near Kitty Hawk on the coast of North Carolina. Rather than improving the prospects for the helicopter, this, if anything, had the effect of leaving it behind, neglected by many pioneers now that the fixed-wing aeroplane had shown the way into the

14

air, encouraging many to join the pursuit of ever larger and faster aircraft.

There were still a few who continued to take an interest in the helicopter, however, even though the attention tended at first to be spasmodic.

In France, Louis Breguet and Richet built a full-sized helicopter test rig at Douai and, on 29th September 1907, this actually succeeded in making a tentative man-carrying tethered flight, powered by a 50-hp Antoinette engine, which drove four rotors. Later, that same year, on 13th November, and still in France, at Lisieux, Paul Cornu actually flew a helicopter with two laterally offset rotor blades, in untethered man-carrying flight, with an Antoinette 24-hp engine for propulsion.

Breguet then concentrated on conventional aircraft, and it was to be almost thirty years before he returned to take up his interest in the helicopter again. He was not the only pioneer to, in effect, pick up the helicopter and put it down again, before eventually returning to it. The talented young Russian designer, Igor Sikorsky, built two helicopters at Kiev in 1909 and 1910, using coaxial rotors, but although lift was achieved, it was still not enough for practical sustained flight, and Sikorsky then dropped the helicopter to concentrate on other more immediately successful types of aircraft.

Strangely, the military neglected the aeroplane at first. This was particularly true of the United States Army, which had developed what can only be described as a blind spot as far as the aeroplane was concerned. This was possibly because it had burnt its fingers, and even made itself seem ridiculous, in supporting Samuel Langley's infamous and unsuccessful rival to the Wright Flyer 1, the Aerodrome A, but the French also simply refused to believe that the Americans had overcome the problems of powered heavier-than-air flight. Only the British Army displayed an early appreciation of the Wright brothers' achievements, and even this interest was lost while the manufacturer and potential customer argued over the question of demonstrations and patents!

Nevertheless, by 1908, the Wright brothers had achieved unstinted recognition and the aeroplane had found acceptance, with demonstrations in the United States and in France, after which its uses for communications and aerial observation or reconnaissance duties were so obvious that a number of countries were soon afterwards operating a handful of aeroplanes each. The more advanced were also conducting experiments into aerial photography and weaponry, while the first flights to and from warships at anchor were made in the

15

United States and then in Great Britain. Regarding this last development, one is tempted to consider that, had the aeroplane followed the helicopter, rather than the other way round, naval aviation as it has developed might never have been seriously considered, bearing in mind the limitations of the early aeroplanes and the very real hazards of flying onto and taking off from ships safely in harbour, which in turn faded into insignificance compared to the hazards awaiting the early aviators when operating from vessels under way. Early naval aircraft confined themselves to just those applications for which the helicopter has proved to be so successful, and at speeds of some 40–70 mph, while the interest in operating land-planes from warships was a direct result of the major disadvantages of having warships stop to put down and pick up seaplanes. The introduction of first the wheeled trolley and then the catapult eliminated the need to stop to launch seaplanes, but for some time afterwards it was still necessary to stop when recovering an aircraft, leaving the mother ship vulnerable to attack and making it difficult for her to keep up with the rest of the fleet. It was only later that ship-borne aircraft obtained a major strike and fighter defence role.

During the 1912 Royal Review of the Royal Navy, off the south coast of England near Portland, four frail aircraft stole the show, demonstrating some of the uses to which they could be put. One aircraft even succeeded in making the first recorded detection of a submerged submarine from the air, although at a relatively shallow depth and by visual observation.

It took slightly longer for the bomber and the fighter to develop and then become firmly established, by which time air speeds had risen to a more respectable level, generally above the at first magical 100 mph figure, for, during its first ten years or so, the aeroplane remained slower than the then fastest steam railway locomotive. Lifting capability increased too, so that before the outbreak of World War I, as many as a dozen people could be carried into the air on one of Igor Sikorsky's "airbuses", and many more were carried by other less brilliant pioneers on publicity and record-seeking stunt flights, often taking a large number of children to increase the head count without adversely affecting the weight lifted! The helicopter obviously played no part in the war, but it is interesting to speculate on what might have been had the crippling German submarine menace in the North Atlantic been countered by escort vessels with anti-submarine helicopters aboard, or had the advancing lines of the then novel British tanks on the Western Front been confronted by missile-armed German anti-tank helicopters, followed by troop-carrying assault helicopters perhaps.

Development, or rather research and experiment, did not cease altogether during the Great War, as World War I was known until some time after the outbreak of World War II. A Lieutenant Petroczy of the Austro-Hungarian Army designed a captive, or tethered helicopter, in April 1915, which he intended to be powered by a 200-hp Daimler engine, but it is believed that this was never built.

There were other pioneers meanwhile, whose interest in the helicopter had been more consistent and less fickle than that of Breguet and Sikorsky at this time. In the United States, Henry Berliner, who had worked largely unsuccessfully on helicopters from 1905 onwards, finally in 1922 produced a machine capable of making a tethered flight for the benefit of the United States Army, but this was still far from being a practical helicopter.

The major problem confronting the early experimenters lay in their failure to appreciate the control problems inherent in the rotary-wing aircraft. The rotor is more akin to the wing of an aeroplane, hence the term "rotary-wing aircraft" than to the airscrew or propeller of conventional aircraft, providing lift as well as propulsion and, of course, also acting as the major control surface. It follows that during one half of rotation, the rotor blade is travelling faster than the helicopter, or advancing, and so creating a considerable level of lift, while during the other half of rotation, the blade is travelling more slowly than the helicopter, or retreating, so creating proportionately less lift that the other half of the rotor blade. The consequent imbalance would at best simply turn the helicopter over. At a time when aeroplane propellers were fixed pitch, with variable pitch not appearing on production aircraft until the 1930s, it was perhaps not too surprising that the importance of the advancing and retreating rotor blades was not immediately appreciated, and nor was the fact that directional control of the machine had to be through adjustments to the angle of the rotor blades.

Even the solutions were not always recognized as such. In 1904, C. Renard advocated flapping rotor blades, but failed to appreciate that this was the simplest solution to the problems raised by the advancing and retreating rotor, with the advancing proportion rising to dissipate the additional lift and the retreating portion falling. It seems that Renard saw the flapping blade as offering a simplified ornithoptering movement, providing additional lift.

The concept of using the rotor blades for directional control is known as cyclic pitch control, and was first suggested by Crocco in 1906, although it was left to Ellehammer, that unfortunate Dane who

One of the earliest helicopters to fly was this 1925 machine by Otienne Oemichen, although it was far from being a practical helicopter.

enjoyed so little success, actually to demonstrate the use of cyclic pitch control in a tentative helicopter design. Progress of a kind was by now being made, in a haphazard way, and the Argentinian Marquis de P. Pescara demonstrated cyclic pitch control on a number of otherwise unsuccessful helicopters between 1919 and 1925, and also demonstrated the use of auto-rotation of the rotor blades for a safe, if bumpy, landing following engine failure. Etienne Oemichen in France built his Oemichen 2 helicopter, and flew it in May 1925, for one kilometre near Arbonane, but this machine, with four coaxial rotors, was far from being a viable helicopter.

The first practical solution to the problem came with what can be best described as an interim machine, offering a hybrid between the aeroplane and the helicopter, but still unquestionably a rotary-wing aircraft, the gyroplane. Although not strictly a new concept, the gyroplane as a workable machine was the brainchild of the Spaniard, Juan de la Cierva, who later became a naturalized Briton, marketing his gyroplanes as Autogiros.

The first successful autogiro flight was with a Cierva C.4 at Getafe in Spain, in 1922, following earlier experiments by de la Cierva using modified Avro biplanes.

De la Cierva cured the tendency of the rotary wing to roll the aircraft over in flight by adopting and modifying Renard's solution, allowing the blades to flap on a simple hinge, and he had perfected this technique by 1922. The mainplane was eliminated on the autogiro, except for a non-lifting stub to carry the ailerons, while the rotors genuinely fulfilled the role of the wings. Take-off was short, rather than vertical, on the earlier versions, with the rotor blades revolving freely in the slipstream from the conventionally mounted tractor propeller. In this form, the autogiro first flew in its production form on 9th January 1923, with Lieutenant Alejandero Gomez Spencer at the controls, and was subsequently demonstrated at Cuatro Vientos in Spain.

In 1925, de la Cierva moved to England to continue his work there, and it was in England, some years later in 1933, that he produced a semi-vertical take-off "jump start" by powering the rotor blades with drive from the engine for take-off, leaving the blades rotating rapidly in negative pitch as the engine rpm built up, and once full power was achieved, turning the blades to positive pitch, making the machine jump up to thirty feet into the air, after which the transition to forward flight could be made and drive from the engine to the rotor disconnected for normal autogiro flight. This was as near to the helicopter as de la Cierva was to go, and clearly the machine was very much a hybrid, with ailerons and a conventional tractor propeller, but within its limitations, the autogiro worked, and proved itself as a practical concept. For de la Cierva, the concept was an outstanding success, for his object was not so much the attainment of vertical take-off, but of flight free from the dangers of a stall during take-off and the initial climb, and which was a particular hazard during the early years of flying: this the autogiro did achieve.

Within a very short time, autogiros were being produced under licence in England (by Avro as the Rota), France, the United States, by Kellet, Pitcairn and Buhl, although the last mentioned produced just one or two aircraft, and Italy, being sold both to civilian and military users. The machine complemented existing units of short take-off general purpose light aircraft, suffering the same limitations on carrying capacity. There were some notable successes, with Lieutenant Alfred Pick of the United States Navy making the first rotary-wing deck landing in a licence-built Pitcairn Autogiro in 1931. A C.30 Autogiro was used by a Royal Air Force officer, Wing-Commander Brie, for a series of landings aboard a cruiser in 1935.

If the autogiro failed to realize the success which was to await the arrival of the helicopter, it was due to the absence of many of the

The Avro Rota, or licence-built Cierva C.30A Autogiro, of the late 1930s.

essential and unique attributes of the helicopter, while it also appeared at a time when national defence budgets were pared to the bone because of the world-wide depression, and there was little enough to spend even on the most vital 'teeth' elements of national defence, let alone on novelty. Yet, on the plus side, the autogiro offered reliability, a reasonable speed and simplicity, even if performance was limited in other respects.

The helicopter itself had not been completely forgotten. In 1935 and 1936, Louis Breguet flew a twin coaxial helicopter successfully, but failed to develop this into a practical production machine, possibly because at this time the future for the French aircraft industry was uncertain, with nationalization and reorganization being forced upon the manufacturers, and the consequent confusion was seriously to impair the French war effort in 1939.

More promising, was the first practical helicopter, the Focke-Achgelis Fw.61 of 1936, a twin-rotor machine with a conventional power unit and tractor propeller as well, which established the first records for rotary-wing aircraft, flying at speeds up to 76 mph and altitudes up to 11,234 feet, and with a maximum duration of 1 hour 20 minutes. Although an experimental machine, as indeed was the Breguet, the Fw.61 led directly to what was supposed to be the first production helicopter, the Fa.223, which the Luftwaffe intended to fly on behalf of the German Army during World War II. An initial order for 100 Focke-Achgelis Fa.223 Drache, or Dragon, helicopters to be powered by 1,000-hp Bramo Q–3 radial engines was in fact issued, but the first ten production machines, completed during late 1942, were all destroyed by Allied air attack. This was to be the pattern, for as helicopters rolled off the production line later in the war, at a purpose-built factory outside Berlin, all but a few machines were destroyed by heavy bombing. A few machines, no more than a handful, survived the war to be captured by the Allies and used for experiment, and apparently general handling and control of this machine was found to be good. Clearly, it was an opportunity lost due to the fortunes of war, since it seems to have been superior in some respects to the early Sikorsky machines.

An unusual variant on the Focke-Achgelis theme was the Fa.330 rotating-wing kite, with a single rotor, towed by German submarines while operating on the surface as a means of extending their limited horizon and providing extra warning of approaching surface vessels. This was an updating of the observation kite concept, tried with limited success by some armies around the turn of the century.

Another German helicopter design, the lightweight Fl.282, powered by a 150-hp Bramo engine, was effectively stillborn as the war ended.

In spite of the work being conducted in France and Germany, the effort devoted to the helicopter at this time paled into insignificance compared with the effort which, for example, was devoted between the wars to the world air-speed record, or even to the development of larger airliners and flying boats. The nature of naval and military aviation was also changing, with a strategic role in which the helicopter could play little or no part gaining in importance compared with the tactical role for which military and naval aviation had been first conceived, and which had been so much the preserve of light aircraft. The change was reflected in the growing number of air forces, autonomous air arms with little commitment to naval or military needs, but concerned entirely with the pursuit of a strategic fighter, bomber, reconnaissance and, sometimes, transport role. This first happened in the United Kingdom, with the Royal Air Force coming into existence as early as 1918, and last in Denmark; but within national variations, with even the United States lagging well behind in this move, the trend was there.

A number of armies retained a few aircraft for observation duties and for communications work, although in some instances these were operated on behalf of the military by the relatively new air forces. Nazi Germany was a case in point, with the Luftwaffe gaining control of all German service aviation by the outbreak of World War II, against strong but fruitless Army and Navy opposition. Battleships and cruisers carried two or three small aeroplanes, usually seaplanes or

A Focke-Achgelis Fa.330 rotating-wing observation kite, ready to be launched from a German U-boat during World War II.

And it could fly! The Fa.330 in the air, extending the U-boat's horizon.

relatively lightweight types of flying boat or amphibian, for reconnaissance purposes, and sometimes for rescue duties, while even air forces had to devote a small number of such aircraft to this role, although landing the flying boats in rough seas was difficult, dangerous, or, frequently, impossible.

In the continued absence of the helicopter, light aircraft for military use became more specialized as new and purpose-designed machines replaced the old general-purpose biplanes, while the growing sophistication and the higher speeds of combat aircraft also increased the need for specialized army co-operation types. Although the Luftwaffe was due to receive Focke-Achgelis Fa.223 Drache helicopters for use in the field supporting the Army, there were also in Luftwaffe service on these operations such brilliant light aircraft as the Fiesler Storch, a high-wing monoplane with a stalling speed of just 30 mph and a maximum speed of 80 mph! In Britain, the Army used Auster light aircraft, with RAF-operated Westland Lysanders, while in the United States, the Army used Piper, Stinson and Sentinel light aircraft, developed from machines on the civil market. Such aircraft remained important, even in Germany, where the helicopter was in theory an early practical proposition. This was due to a number of reasons, not least being the fact that design, development and the setting up of production for most wartime aircraft took place during the years immediately preceding the outbreak of hostilities, so even the Drache was a couple of years behind this, and of course, the relative simplicity of light aircraft also meant that pilots could be trained more cheaply and quickly than, for example, gyroplane pilots could be. Had the most optimistic of German helicopter production schedules been achieved, with a rate of production of some 400 machines a year, it is doubtful whether sufficient pilots of sufficient skill could have been found, and trained, and if it had been possible, one could only guess at the effects on other areas of the German war effort.

During the war in the Pacific, the United States Marine Corps even had to re-invent integral airpower of a kind. Having developed into an operator of fighter-bombers at an early stage, the USMC suddenly realized during the Solomons Campaign during late 1944 and early 1945, that it had to seek out the ex-private fliers and light-aircraft mechanics from amongst its ranks of infantrymen to find the experience to operate light aircraft, twenty-four ex-US Army Piper O–1 Cubs, in the forward communications and observation roles. The United States Army Air Corps, which became the United States Army Air Force shortly before the United States entered World War II and which eventually became the United States Air Force in 1947, had already been divided from the integral element of US Army aviation, giving this service what amounted to two air arms during World War II!

At sea, the Supermarine Walrus, a successor to the Seagull, performed certain of the duties which were later to fall to the helicopter, operating from warships equipped with catapults. Seaplane carriers, which had been so important during World War I, had lost their more glamorous duties to the aircraft-carrier, but a small number remained in service, even though three Japanese vessels of advanced design were converted to aircraft-carriers, after an interim period as midget-submarine carriers! Escort carriers, either converted merchantmen or based on merchant vessel hull forms, and merchant aircraft-carriers, which continued to carry cargo in addition to a small number of aircraft, carried aircraft such as the veteran Fairey Swordfish biplane to provide anti-submarine patrols for convoys, again a task well suited to the helicopter. However, these vessels sometimes carried fighters as well, or instead of anti-submarine aircraft, to provide air cover.

The need to provide adequate protection for convoys and the value of providing this protection from the air, had been recognized as early as 1913, when Admiral of the Fleet, Lord 'Jacky' Fisher, demanded that the Royal Navy should have anti-submarine airships for coastal patrols. The simple expedient of placing a BE2c aircraft fuselage under the envelope of an airship solved the problem and, by 1915, sixteen such craft were in service with the Royal Naval Air Service. Larger airships, the *Parseval* and the *Astra Torres*, escorted the British Expeditionary Force across the Channel to France on the outbreak of war, while later on, after the convoy system became widespread, airships provided protection for coastal convoys. Generally, airships were shore-based, but some were carried by warships, including HMS *Furious*, the converted battle-cruiser which became the world's first aircraft-carrier, and which carried an airship moored on her quarterdeck.

On the entry of the United States into World War II, some fifty airships were in United States Navy service. Although one, K–74, was shot down by a German U-boat in July 1942, these craft were generally able to operate safely outside the range of German and Japanese aircraft and no surface vessel was lost in an airship-escorted convoy. Later in the war, USN airships were based in Brazil for patrols in the South Atlantic. The Royal Navy did not use airships during WW II.

This was the situation when Igor Sikorsky returned to the development of the helicopter, having fled from his native Russia

A British RNAS coastal airship escorts a North Sea convoy during World War I.

The equivalent of today's helicopter landing platform. The first aircraft-carrier, HMS *Furious*, in her interim form, carries a balloon on the landing deck.

The first practical helicopter, the Vought-Sikorsky VS–300 flying in an interim configuration (note the extra tail rotors) with Igor Sikorsky at the controls.

during the Bolshevik Revolution to live and work in the United States, where he soon established a reputation as one of the leading designers of flying boats and amphibians, with his aircraft pioneering many of America's overseas air routes.

In 1938, he started work on his VS–300 (Vought-Sikorsky 300) helicopter, which he was able to test in tethered flight during the following year, on 14th September, with little immediate success, even though the machine, with the classic configuration of a single main rotor and a small tail rotor, driven by a 75-hp Lycoming engine, incorporated cyclic pitch control in an attempt to achieve the full range of movement. Cyclic pitch was abandoned for a further programme of test flights during 1940, with a 90-hp Franklin engine driving a single main rotor, while the tail rotor was supplemented by two small horizontally-mounted rotors on outriggers, intended to provide improved lateral control. In this form, the VS–300 made its first free flight on 13th May, while almost a year later, on 6th May 1941, it established a world endurance record of 1 hour 32 minutes.

During the remainder of 1941, Sikorsky refined the basic design further, restoring partial cyclic control during the summer, then removing the outriggers and their rotors, and finally, at the end of the year, restoring full cyclic pitch control, installing a 150-hp Franklin engine and replacing the skids of the earlier forms with a tricycle

The Vought-Sikorsky VS–300 in its final form, again with Igor Sikorsky as his own test pilot.

undercarriage. There were also changes to the fuselage, which had originally been fabric covered, but was modified to a stripped down open framework for the 1940 and 1941 experiments, before being re-covered for the final programme of successful test flights during 1942. Even while the final tests were being conducted, work was proceeding on the Sikorsky XR–4, the predecessor of the R–4 series, on which most of the early experiments with helicopters were to be made.

By coincidence, Sikorsky's success coincided with renewed interest in the helicopter amongst airframe designers, and work on a different approach to the problems of the practical helicopter was also taking place in Austria, for example, where Doblhoff's No. 1 helicopter was tested in 1943, featuring a rotor with jets in the blade tips. Others taking an interest in the helicopter at this time included the Kellett Autogiro Company, with an interesting design using twin interconnecting rotors, and Landgraf.

However, out of all the different methods proposed, it was the single rotor helicopter which was to have the greatest impact, offering a convenient size with the minimum complication and maximum manoeuvrability. It was soon given a chance to prove itself.

2
WORLD WAR II
OPERATIONS AND EXPERIMENTS

The United States Army had been aware of the potential of the helicopter for many years before the development and success of the Sikorsky VS–300. As early as 1918, the Army's Air Service Engineering Division at Mount Cook Field investigated a helicopter design by Peter Cooper Hewitt, and drew attention to the very great potential of vertical take-off flight for restricted landing areas. Just a year later, there were tests with a quarter-scale model of the McWhirter "Autoplane".

By coincidence, if one remembers Igor Sikorsky's Russian parentage and upbringing, it was to another Russian *émigré* that the first United States Army contract for a helicopter was given. George de Bothezat had originally approached the US Army in 1919, and in 1921 he was awarded a contract for the construction of a helicopter, worth a total of $20,000, but payable in instalments and on completion, with a number of contractual stipulations, the last of which was a fully successful flight! Possibly this stringency was due to still fresh memories of the costly failure of Samuel Langley's "Aerodrome A" in 1903. De Bothezat had completed his strange craft, with four six-bladed rotors mounted on diagonal cross members and with the sole occupant, the pilot, sitting in the cross-over of the diagonals, by late the following year, ready for Army trials at Mount Cook Field. The first flight came on 18th December, when, at 09.00, the machine rose smoothly and without undue fuss to the height of six feet, where it remained for almost two minutes, moving with the airstream for about 300 feet before the pilot, a Major Bane, descended and then ran into the perimeter fence.

After this encouraging start, the de Bothezat managed to lift two men off the ground on 19th January 1923, but although the original 180-hp Le Rhone rotary engine was later replaced by a more powerful 220-hp unit, the helicopter remained a purely experimental and impractical machine due to the limitations of its performance. Controls on the de Bothezat were similar to those of an aeroplane, with stick and pedals, but with the addition of a small wheel to control the cyclic pitch of its rotor blades.

However, limited though the performance of the de Bothezat was, it was an encouraging start, and the Army's pursuit of the helicopter continued. During the 1930s, the scene for the experiments of the day was the National Advisory Committee on Aeronautics' Langley Memorial Aeronautical Laboratory, at Langley Field, Virginia. Here, in 1936, NACA tested two licence-built autogiros, a Kellett YG–1 and a Pitcairn YG–2, although the latter crashed after the canvas covering on the rotor blade ripped loose in flight. Subsequently, the Kellett YG–1B was put into limited production and the newly formed US Army Air Corps Autogiro School at Patterson Field, Fairfield, Ohio, operated these machines from its inception on 15th April 1938.

The pace of development quickened somewhat with the approach of a new decade. Laurence Lepage and Havilland Platt had visited Germany to look at the Focke-Achgelis Fw.61, and on their return had designed the Platt Lepage PL–3 helicopter for the United States Army. Development was delayed, however, until on 30th June 1938, Congress authorized the expenditure of the then not inconsiderable sum of $2 million on research into rotary-wing aircraft. An order for the PL–3, which was redesignated XR–1 by the US Army, was placed on 19th July 1940. The machine, not surprisingly, owed much to the Focke-Achgelis Fw.61, with an aeroplane fuselage with wings which had a high rate of anhedral and which were fitted with wingtip-mounted rotors driven by a Pratt and Whitney R–985 450-hp engine. The machine provided tandem seating for two in its cockpit, while a Perspex nose ensured an excellent forward view. Tethered flights of the XR–1 started on 12th May 1941, and subsequently the aircraft was joined by a developed version, the XR–1A, but early flights proved to be disappointing, and it was not until spring, 1943, that the XR–1 could make the first helicopter flight over a closed circuit, and not until December of that year that an altitude of 300 feet could be attained, raised in 1944 to 600 feet.

Even these limited results passed for success at the time, but wisely the US Army had decided to support a second, rival programme, that of Igor Sikorsky, with the VS–300, mainly to gain any benefits which might accrue from parallel development of a different design philosophy. At the time, few could have predicted that it was to be the Sikorsky single-rotor design which offered the best chance of success, but as early as December 1940, the US Army decided to order two Vought-Sikorsky XR–4 helicopters, known to the manufacturer as the VS–316.

The Army's use of XR designations for all rotary-wing aircraft at the time is confusing. The XR–2 and XR–3 were not in fact helicopters, but improved autogiros, with the XR–2 being a "jump-start" machine from the Kellett Autogiro Corporation, and the XR–3 a cyclic pitch control autogiro from the same company, and both of which were developments of the earlier YG–1B in the same way that Cierva himself developed his earlier machines and offered a "jump-start" variant.

At first, the XR–4 was planned to follow the interim configuration of the VS–300, with a single main rotor and three small auxiliary tail rotors, two of which would have been horizontal. However, there were many changes during the development of these machines while the VS–300 continued its programme of trials, and eventually the XR–4s were completed with just one vertically mounted tail rotor as the result of the successful trials with the VS–300 in this form. Indeed, it was found that the loss of the two small horizontally mounted tail rotors actually improved performance. Unlike the single-seat VS–300, the XR–4 was a side-by-side, two-seat machine, offering a reasonable view forward for the two crew members, while the US Army specification called for a 550-lb load-carrying capability. It was to be a workable, and worthwhile, helicopter in many respects, even though it used just one small 165-hp Warner engine, mounted immediately behind the cabin.

After being wheeled out just before Christmas 1941, the first flight of the XR–4 followed on 14th January 1942, with Sikorsky's Chief Test Pilot, Les Morris, at the controls. Morris flew the helicopter for a total of six flights that day, remaining in the air for some 25 minutes altogether, with the longest individual flight being for 7 minutes 20 seconds. This in itself was a very good start, with a first day's flying time well in excess of the total airborne time for many of the earlier experimental helicopters.

As the year progressed, even better was to follow. On 20th April 1942, Les Morris flew the XR–4 on a demonstration arranged for representatives of the British and American armed forces in a field not far from the Sikorsky works at Stratford, Connecticut. The XR–4 was flown to an altitude of seven feet, hovered, and then dropped back onto the take-off spot, before rising again to fly sideways, backwards, up and down, and then, of course, forward. A small hoop had been placed on top of an eight-foot-high post, and the XR–4 hovered in front of this post, before edging forward to spear the hoop with its air-speed indicator pitot tube, and then finally flying with the hoop to hover in front of Igor Sikorsky so that the great inventor could take the hoop off the pitot tube. For the assembled spectators, who included Wing-Commander R. A. C. Brie, RAF, and Commander J. H. Miller, RN, this was a revelation. More demonstrations followed that day, with the XR–4 demonstrating its ability to land and pick up personnel or casualties, and the VS–300 appearing, with inflated flotation bags, to land on the Housatonic River, and then take off again.

However, all this good work had been conducted by the manufacturer, and the US Army naturally wanted to have its new aeroplane for trials at its main experimental establishment at Wright Field, Dayton, Ohio. Logic dictated that the XR–4 be dismantled and taken by road, but the need to demonstrate and test remained, and so the courageous decision was taken to fly the machine over the 760-mile route from Stratford to Dayton, a more direct flight of 560 miles having been wisely rejected because of high mountains across the route.

Les Morris was once again in the pilot's seat when the XR–4 took off on the longest helicopter flight at that time, on the morning of 13th May 1942. Throughout the flight, the XR–4 was accompanied by a car with a team of Sikorsky engineers, although at times the frail and underpowered helicopter was hard-pressed to follow the car, running into 15-mph head winds which reduced the XR–4's 60-mph cruising speed to 45 mph over the ground, but generally, the impression given to the pilots was that it was a good flying machine at low altitude, even in turbulence. Igor Sikorsky himself piloted the XR–4 over the Cleveland to Mansfield section. Altogether, the flight was spread over five days and took 16 hours 10 minutes, with sixteen stops, only eleven of which had been scheduled originally. Even so, this was a record for rotary-wing aircraft, and there were no serious mechanical hitches, although the gearbox showed occasional signs of overheating.

The flight test trials at Wright Field were not confined to Army tests alone. On 7th July 1942, the XR–4 was demonstrated for General Arnold, Chief of Staff of the United States Army Air Force, the direct predecessor of the USAF, and before long five USAAF officers were receiving helicopter flying training. The XR–4 was used on trials to evaluate its usefulness as an anti-submarine aircraft, able to fly off merchant vessels, dropping 25-lb bombs carried on the lap of a passenger while flying over the outline of a submarine. Later, doubtless to the relief of these passengers, racks for up to eight 25-lb bombs were fitted under the fuselage, but oddly enough, pilots found that the most accurate bomb delivery occurred while flying at 40 mph rather than when hovering!

The trials continued throughout 1942. During the autumn, inflatable flotation gear was fitted, with some early problems with vibration when landing the float-equipped XR–4 on hard surfaces, although these difficulties were soon overcome. An altitude record of 12,000 feet was reached before the first series of trials ended in January

1943, and the YR–4 helicopter was ordered into production.

A more thorough and demanding evaluation of the shipboard use of the XR–4 was soon to follow. The ss *Bunker Hill*, a tanker owned by the United States War Shipping Administration, was prepared for deck-landing trials of the new helicopter. Although the value of a stern landing platform was clearly understood, even at this early stage, the layout of this particular tanker did not allow this, and instead a wooden platform was fitted aft of the superstructure, but forward of one of the masts. This meant that the landing approach had to be from abeam, with only a fourteen-feet clearance between the rotor-blade tips and the superstructure forward of the platform or the masts aft of it. This was difficult, not least because of the very real problem for the pilots in seeing the mast stays, and it was also unnerving for the spectators who, as the helicopters flew sideways onto the ship, found themselves on the decks of the superstructure in a fine position to be decapitated by the whirling rotor blades! However, there were to be no beheadings during the trials, or afterwards!

The trials started well on 6th May 1943, while the ship lay at anchor in Long Island Sound, giving the pilots time to adjust to their awkward landing site. Later that day, demonstrations began, with a total of twelve landings, both at anchor and then later while the ship was at sea, operating at varying speeds. Little difficulty was found in landing while the ship steamed at speeds of up to 7½ knots, but above this, increasing problems were found with turbulence created by the flow of air around the ship's superstructure. On the second day, flying had to be delayed until after a thick mist began to lift at 10.30, and eventually started with visibility at two miles. Most of the second day's flying was by USAAF pilots, including Colonel Frank Gregory, who had been first to land on the ship. These trials finally persuaded the United States Navy of the value of the helicopter in operating from ships at sea. A final proof of the value of the helicopter came at the end of the trials, when the XR–4 carried the Chief of the United States Army Air Office's Development Branch, Colonel R. C. Wilson, from the ss *Bunker Hill* direct to Stratford Airport so that he could board his plane to Washington.

Further trials followed, this time aboard the ss *James Parker*, a liner converted to a troopship for the United States Army Transportation Corps, with the specific aim of assessing the value of the helicopter as a shipboard anti-submarine weapon carrier. A stern landing platform was available for these trials, which started on 5th July, using both an XR–4 helicopter with a conventional wheeled undercarriage and one of the first YR–4 helicopters, fitted with flotation bags. The initial flying trials took place while the ship lay at anchor, with Lieutenant-Colonel Les Cooper flying the XR–4 and Lieutenant Frank Peterson flying the YR–4B for landing and take-off trials, with the XR–4 landing first and being quickly pushed out of the way to allow the YR–4B to land. Sikorsky's Les Morris and three additional US Army pilots joined the ship for the trials, at anchor on 5th and 6th July, and under way on 7th and 8th July. Again, turbulence was encountered as the ship's speed increased, but less difficulty was caused due to the positioning of the landing-platform at the stern. The trials included flying into a stiff breeze and, on 7th July, the helicopters continued to fly while the ship rolled through ten degrees, although later that same day flying had to be temporarily suspended after the weather worsened and heavy rainstorms were encountered. Altogether, over the four days, the two helicopters were airborne for a total of 20 hours, divided almost equally between them, and made a total of 162 deck landings and take-offs, with and without passengers.

The success was particularly surprising because the trials had been in real danger of cancellation at the last minute. While the final preparations were in hand, an Army YR–4, flown by one of the first United States Coast Guard pilots, Commander Frank Erickson, was involved in an accident while still at the Sikorsky works at Stratford, with the main rotor blades striking the tailplane and causing the helicopter to crash-land, although without causing any injury to Erickson. It was decided to press ahead with the trials, while modifications, increasing the clearance between the rotor blades and the fuselage, were incorporated in the YR–4B. Other alterations to the YR–4B included fitting a 180-hp Warner engine, larger 19-foot rotor blades, and a radio – the early VS–300 and XR–4 pilots had flown without radio contact with those on the ground.

The YR–4B underwent certain trials of its own, notably the cold-weather trials in the Arctic. On 6th November 1943, Lieutenant-Colonel Les Cooper, US Army, collected a YR–4B, which he flew to a USAAF base for dismantling, so that the machine could be loaded into a Curtiss C–46 Commando transport for the long flight to the USAAF's cold weather research unit at Ladd Field, near Fairbanks in Alaska. The exercise also tested the air portability of the YR–4B, which had its rotor and tail rotor removed, to be suspended from the ceiling of the C–46, while the cabin, tailplane and undercarriage were also separated for the flight. On arrival at Ladd Field, the YR–4B, nicknamed 'Arctic Jitterbug' was reassembled and prepared for flying, using cold-weather grade lubricants, and was put onto a test programme, which included air-ambulance and liaison duties. For the air-ambulance operations, the first external stretcher structure was designed and built, with canvas covers so that 'guinea pig' patients could be flown while inside sleeping-bags – with surprisingly few complaints! Overall, the Arctic trials were a success, and little effect was noted on the stability of the YR–4B with its external human loads. While careful attention had to be paid to the problem of icing of the rotor blades while on the ground, there was little difficulty over icing while in flight.

Elsewhere, another YR–4B was used for evaluation by the US Army Signal Corps at Fort Monmouth, New Jersey. A demonstration of the YR–4B was also arranged to take place at the Capitol in Washington, DC, which led to Colonel Frank Gregory flying the small aircraft into Washington on 16th May 1943, in thick early morning fog, although he still managed to arrive some fifteen minutes early! Shortly after his arrival at the Capitol, a further flight was made, carrying mail from the Capitol to Washington's Municipal Airport. That afternoon, a demonstration of the YR–4B's capabilities was staged for the President of the United States.

Meanwhile, the YR–4B was steadily evolving from having been a trials and demonstration helicopter, proving the worth of this newest type of aeroplane, to being a training helicopter. The United States Coast Guard had been amongst the earliest converts to the helicopter, and in November 1943, three YR–4Bs, by this time known to the USCG and the United States Navy, as the SMS, were borrowed from the USN and installed at Floyd Bennett Field at Brooklyn, New York, which was then designated as the official USCG helicopter training station.

It was not long before these helicopters were able to prove their operational value, in earnest! On 31st December 1943, there was a sudden explosion aboard an American destroyer, the USS *Turner*, while on station off the coast of New Jersey, resulting in very many casualties. The immediate rescue of the ship's company, who were taken ashore, was made by USCG vessels, while Commander Erickson flew one of the YR–4Bs from Floyd Bennett Field to Battery Park, New York, to pick up vital and urgently required blood plasma, which he then flew direct to the first-aid stations providing emergency treatment for the crew.

The United States Navy itself had reported favourably to the Navy Bureau of Aeronautics on the advantages of the helicopter, including its value as an anti-submarine weapon. The Navy had been particularly impressed by the small size and manoeuvrability of the helicopter, compared with the airship, which had been used as an anti-submarine convoy escort, but which also suffered from handling difficulties while on the ground or aboard ship. It also seemed likely even at this early stage that the helicopter could cover a far wider area than could any one escort vessel or airship in a given time, being faster than either, and yet also having a lower fuel consumption. Following the success which had attended the trials aboard the USS *Bunker Hill* and the SS *James Parker*, two further vessels were fitted with landing-platforms, the USCG vessel *Governor Cobb* and the British SS *Dagheston*. Sea-going trials started aboard the *Dagheston* on 28th November 1943, with USCG and USN pilots flying specially modified YR–4s fitted with floats, making 166 deck landings aboard the *Dagheston*, followed by a further 162 landings with Royal Navy pilots at the controls. By the end of 1943, further trials had taken place, and these continued throughout 1944, with two Sikorsky YR–4B helicopters embarked aboard the *Dagheston*, with five Royal Navy pilots, one pilot from the USCG, and four more from the USN.

The Royal Navy pilots in these experiments had been drawn from the newly formed Helicopter Service Trials Unit, while further pilots were trained on behalf of the British Admiralty by the USCG at Floyd Bennett Field. The Royal Navy allocated four helicopters on which the USCG could train British personnel, and altogether more than 100 Royal Navy pilots and 150 mechanics were trained for subsequent helicopter operations by the Royal Navy. The RN helicopters used for the USCG training programme were mainly the improved production versions of the earlier XR–4 and YR–4B series, the more powerful R–4Bs, fitted with a 200-hp engine and offering increased range. The Royal Navy received most of the forty-five R–4Bs supplied to the British armed forces under America's generous wartime Lend-Lease Scheme, with these helicopters being known in Britain as the Sikorsky Hoverfly.

The Royal Air Force also used a small number of R–4Bs and, in August 1944, started to replace the Avro-built Cierva C–30 and C–40 Rota autogiros of No. 529 Squadron with the R–4Bs, while additional examples were supplied to the new RAF Helicopter Training School near Andover during early 1945. In RAF service, the R–4Bs also undertook radar calibration work, and one machine even at this early stage joined the King's Flight, the RAF's main VIP unit specializing in carrying members of the British Royal Family and, more usually, British Government ministers, although the Hoverfly's role with this unit was confined to carrying urgent mail and dispatches rather than any VIPs!

The new RAF helicopter squadron, No. 529, had itself been formed

The first operational helicopter, a Sikorsky R–4 of the USAAF hovers above an R–5 model.

only on 15th June 1943, from No. 1448 Flight.

Altogether, some 100 R–4Bs were ordered by the USAAF, but twenty-two of these were passed to the United States Navy and Coast Guard without entering USAAF service, leaving the USAAF as the procurement agency for the helicopter at this early stage. In USAAF service, these machines were designated the HNS–1.

While the pre-production YR–4B had travelled north to Alaska for Arctic trials, the R–4B was sent, if not exactly in the opposite direction, at least to a far warmer climate, to Burma for tropical operating trials. It was in Burma that the helicopter was used on its one and only World War II operational role when, in March 1944, Lieutenant Carter Harman flew an R–4B from India, over a 5,000-feet mountain range, to the Allied base at 'Aberdeen', in Burma and behind Japanese lines. The trip itself was an achievement, without the added danger of being in a war zone, being flown with extra fuel tanks strapped to the machine above the head of the pilot to increase the range by an extra ten miles beyond the R–4B's usual 120 miles. 'Aberdeen' was a secret base occupied by Allied bomber and transport units, and by the British 1st Commando Group, and it was from here that the little R–4B was to rescue three British soldiers and a downed pilot, flying them out one at a time from a paddy-field.

36

Successor to the R–4, the Sikorsky R–5 offered vastly improved forward and downward visibility, although accommodation was limited.

The YR–4 and the production R–4B had established the helicopter in an unassailable position as a vital piece of military and naval equipment. There could be little doubt about the helicopter's ability to handle rescue, communications, liaison and casualty-evacuation duties, as well as anti-submarine warfare, spotting submarines from the air, at sea and well beyond the range of shore-based maritime-reconnaissance aircraft, and without the need for support from scarce and costly aircraft-carriers. No one who came into contact with the machine, even while still in its infancy, could fail to be impressed by the possibilities of this still novel form of flight. However, there were still severe limitations in performance, and it was these problems which the manufacturer and the US Army now started to rectify, designing helicopters with specific purposes in mind, and to meet particular minimum performance specifications.

Early in 1943, Sikorsky started work on a new helicopter, designated the Vought-Sikorsky VS–327, to meet the Army's requirements, and incorporating several features to ensure that the customer received a better observation helicopter. Larger than the R–4 series, the VS–327 bore little resemblance to its predecessor, other than conforming to the by now classic single-rotor configuration. The new helicopter had a more streamlined shape with extensive use of

37

Plexiglass in the nose to provide an outstanding all-round view, an effect which was emphasized further by the inverted 'pear' shape of the fuselage cross-section, and by placing the pilot behind the observer in the tandem layout two-seat cabin. Designated the XR–5, the USAAF originally ordered two prototypes at a total contract cost of $650,000, but this was soon increased to four aircraft, at a cost of $1 million, with the extra two aircraft being intended for Lend-Lease to Britain.

It is sometimes forgotten that even military aircraft construction suffered from shortages of materials during World War II. This same shortage of materials had, earlier in the war, enhanced the appeal of the de Havilland Mosquito fighter-bomber to the Royal Air Force, and it even stretched across the North Atlantic to the United States aircraft industry, in spite of the distance from the war zone of that country. This was partly because the American aircraft industry was struggling not only to supply its own armed forces, but also to a great extent, those of its Allies, including Australia and New Zealand, cut off from their normal supply of British aircraft, and even to top up British aircraft production. This shortage of materials had an adverse affect on the development period for the XR–5, and on the design itself, forcing the manufacturer to abandon the use of aluminium for the tail structure in favour of plastic-impregnated plywood. The rotor of the VS–327, or XR–5, used a laminated wood spar, wooden ribs and fabric-covered trailing edges, while the aircraft was powered by a nine-cylinder 450-hp Pratt and Whitney R–985–ANS Wasp Junior engine.

Development flying started with the XR–5 on 18th August 1943, with Les Morris at the controls, taking the prototype to just one foot above the ground, as usual this was at the Sikorsky plant at Stratford. The flight was a disappointment, its success being marred by severe vibration from the rotor blades. Nevertheless, by the end of the month, the XR–5 was in the air again, flying for a total of 33½ minutes, with the longest individual flight being one of nine minutes. Clearly, at even this early stage in its development, the teething troubles were behind the XR–5, and if further evidence of this were needed, on 13th September 1943, the machine lifted ten people, eight of whom were hanging on outside, to an altitude of 800 feet, and also managed a speed of 75 mph! There were in fact a few further problems to be resolved before production of the pre-service YR–5 could begin, but this did not take long, and the USAAF received 26 YR–5As, followed by a further 34 of the production R–5A. Eventually, five YR–5As were converted to dual control YR–5Es for duty as training machines, and later many of the R–5As were converted to carry a third occupant, as R–5Es, using 600-hp Pratt and Whitney R–1340 engines, a nosewheel undercarriage instead of the tailwheel of the XR–5 and YR–5A, and a rescue hoist.

Two of the USAAF's small order for R–5As were passed to the US Navy, which operated these machines with the designation HO2S–1. If USN interest seemed to be lukewarm at this time compared to that of the Army, it was not due to lack of enthusiasm or appreciation of the helicopter's merits, but simply to the lack of suitable ships from which to operate helicopters, while the limited performance of the early helicopters made true anti-submarine work or effective plane guard duties from aircraft-carriers difficult and unsatisfactory.

More tangible evidence of the United States Navy's real enthusiasm for the helicopter was not long in coming. Although the R–5 series was a development of the R–4, a refined and developed R–4 was also considered worth while, and so Sikorsky went to work to develop the VS–316B, which was given the designation XR–6 by the US Army Air Force. Unfortunately, having started the project, the USAAF had second thoughts about it, and it was only the intervention of the US Navy, exerting pressure for a helicopter with greater performance than the R–4B, which saved the project. The USN initially proposed to buy four aircraft, but in due course, of the five XR–6 prototypes built, three passed to the USN and two were taken by the US Army Air Force. The twin-seat, side-by-side layout of the R–4 was retained, with the rotor and transmission of the earlier helicopter, but although a 225-hp Lycoming engine was at first envisaged, this was replaced on the prototype machine by a 245-hp Franklin six-cylinder engine. At the design stage, provision was made to carry stretchers on either side of the fuselage, while bomb racks could be fitted underneath the fuselage.

The XR–6, in spite of the use of many well-proven components, also suffered early problems, perhaps not surprisingly due to the still elementary stage of helicopter development at the time, and when Les Morris took the first prototype into the air on 15th October 1943, he found the controls to be stiff, with a heavy down load on the control lift levers. By the 27th November, the performance was still far from satisfactory, although he could keep the helicopter in the air for several minutes at a time, and it was not until the following January that the problems were resolved. However, after this uncertain start, the new helicopter was soon able to prove itself, setting a new record for a non-stop helicopter flight on 2nd March 1944, when Ralph Alex flew 387 miles from Washington National Airport to Patterson Field, Ohio, in 4 hours 55 minutes, carrying a passenger and also facing head winds of between 10 and 30 mph. Two other records set during this flight

The Sikorsky R–6 attempted to combine the best features of the R–4 and R–5, with side-by-side seating and a bubble canopy.

included the longest continuous period in the air for a helicopter at this time, and an unofficial helicopter speed record. *En route*, the helicopter had to climb over the Allegheny Mountains, rising to 5,000 feet at one point.

Five XR–6As followed, with two for the USAAF and three for the USN, before a pre-production run of 26 YR–6As was built by the Nash-Kelvinator Corporation, which was subsequently responsible for producing some 200 R–6As during 1945. At this time, Vought-Sikorsky was fully committed and, of course, the company lacked the helicopter production capacity which it was to develop in later years. Of the production machines, some forty R–6As were delivered to the USN as HOS–1s, to complement the earlier three XHOS–1s or XR–6As, and these were the first helicopters to reach naval squadrons, with the first squadron being VX–3, commissioning in July 1946, and undertaking air-sea rescue and liaison and communications duties. The United States Coast Guard also received a small number of machines. Some thirty R–6As passed to Great Britain under the Lend-Lease Scheme, being designated the Hoverfly II, with the Royal Navy using just a couple of these machines and most of the remainder entering service with the Royal Air Force, which initially assigned them to the 657 (AOP) Squadron in support of the Army, and to the Airborne Forces Experimental Establishment.

The end of World War II naturally did not mean the end of

helicopter development and experiments by the military and by the leading navies. Indeed, the aircraft designers of the victorious nations were soon provided with captured German helicopters and even captured German designs, which helped to develop and clarify their own ideas, while the leading German expert, Professor Heinrich Focke, went to work in France for the SNCA du Sud-Est, one of the predecessors of Aerospatiale.

Focke had, as mentioned earlier, followed his successful Fw.61, a coaxial twin-rotor helicopter with a tractor propeller and bearing some resemblance to the gyroplane, with the Fa.223 Focke-Achgelis Drache or Dragon. The twin coaxial rotors were about all of the resemblance left between the two machines. The Fa.223 had featured an amidships-mounted engine, that is behind the cabin as on the Sikorsky machines, with which it also shared fabric-covered steel tube construction. However, there were differences in the designs, with the Fa.223 providing a four-seat cabin with a Plexiglass nose offering superb visibility forward, and also used a much larger power-plant in the form of the 1,000-hp Bramo 323 radial engine. An interesting design between the Fw.61 and the Fa.223, was the Fa.266 Hormise, of which three prototypes were to have been built, although none of these seems to have been completed, and attention was concentrated on the Fa.223 series. Reputedly, general handling of this advanced machine was excellent, and a helicopter altitude record of 23,294 feet was established on 28th October 1940. Early in 1942, after a second prototype had been built, the type was ordered into production and, as already mentioned, suffered heavily from effective Allied bombing. On Germany's surrender in 1945, only three Fa.223s remained in airworthy condition, although another three were completed after the war using salvaged parts, of these one was built in France by Heinrich Focke himself, and two were built at the Czech Avia factory. One of the three wartime machines was acquired by the Royal Air Force and flown to the Airborne Forces Experimental Establishment in September 1945, and in so doing became the first helicopter to fly across the English Channel. It did not last long for, sadly, after its historic achievement, it crashed on its third flight in Britain.

Had Germany not lost the war, helicopter development might have taken a different course, but it is unlikely. Germany's declining military position and the gradual loss of control of her own airspace from 1942 onwards meant that the production of many promising machines was hindered and eventually made impossible by increasingly accurate Allied air attack on the factories. The real difference must be that the World War II active service record of the helicopter was far less than it might have been had the Germans managed to deploy their helicopters on the battlefields and at sea, which they almost certainly would have done, given their closer proximity to the war zone than the Americans. Post-war development of the German designs elsewhere in Europe did not show them as offering a suitable rival line of development to that pursued by Sikorsky and some of his American competitors, although it may be argued that the French and Czech designers who were the first to take up Focke's work lacked his talent. Focke himself did design other helicopters, including a large four-rotor machine, effectively a marriage of two Fa.223s, and a large Fa.284 helicopter crane, to be powered by two BMW 2,000-hp engines and designed to lift a load of some 15,000 lbs – approximately seven tons!

With so much of their aeronautical talent having fled from the excesses of the Bolshevik Revolution, the Soviet Union lagged behind the Americans and the Europeans in aircraft development during the 1920s and 1930s. However, what can only be described as the Russian 'feel' for the helicopter still managed to manifest itself at an early stage, although in common with some of the designers on the other side of the Atlantic, the Russians tended to be heavily influenced by the German Fw.61. The first design to evolve within the Soviet Union, the Omega, also used the coaxial twin-rotor system, with the rotors mounted on fuselage outriggers, although the designer, Ivan Bratukhin, adopted the unusual step of mounting the helicopter's twin engines with the rotors on the outriggers. The helicopter actually entered production, with the prototypes using Russian MV–6 engines, pre-production models used American-supplied Pratt and Whitney R–985–AN–1 engines, similar to those used in the R–5 series, and production machines used 500-hp Ivchenko AI–264R engines.

The first flight of the prototype Bratukhin Omega 2MG did not occur until early 1943, suggesting that there may have been some development problems after the machine was completed in mid-1941, and it appears that the 220-hp power-plants of the prototype left much to be desired both in terms of reliability and power output. A more powerful machine with 350-hp MG–41F engines was available by 1944, although not displayed publicly until 1946. The later pre-production and production machines were built and flown up to about 1950, when the Bratukhin designs were abandoned, the design team dissolved, and the last of the Focke-Achgelis related designs was consigned to the scrap-heap.

Of course, significant progress was still being made elsewhere, and particularly in the United States as other companies hurried to join the growing number of helicopter manufacturers, realizing that the new type of aeroplane offered a lucrative market. Unlike the early days of aviation, the helicopter arrived at a time when the industry was well-established, and most of the newcomers to the helicopter industry were companies already well-established as manufacturers of fixed-wing aircraft. Sikorsky, for example, had a first-class reputation as a builder of flying boats and amphibians, while Bell, one of the next companies to enter the helicopter field, had a reputation as a builder of fighter aircraft often of unusual configuration. Bell's first helicopter was the experimental Bell Model 30, on which the company introduced its novel twin-bladed rotor. The Bell rotor system was in fact more akin to a variable pitch propeller, with the blades not independently articulated, but instead worked together with control exercised by rocking the twin-bladed rotor around its hub to produce an effect similar to that of cyclic pitch. Running at right angles to the rotor blades, the Bell system incorporated a short stabilizing bar.

Five Bell Model 30s were built, with the first of these flying during early 1943, and all used 165-hp Franklin engines. At first the 30 suffered from excessive vibration, but this was cured by early 1944. A small twin-seat side-by-side cockpit or cabin in the 30, and the machine's nose, had an appearance similar to that of a light aircraft rather than a helicopter, but from this was to be developed one of the most distinctive and most successful helicopters ever, the Bell 47, which first flew on 8th December 1945, using a 178-hp Franklin engine.

Also in the United States, Frank Piasecki, one of the engineers who had worked on the Platt-Le Page XR–1 helicopter, started his own design team to develop a single-rotor helicopter, in sharp contrast to the twin rotor designs for which he was eventually to become famous. Piasecki's company, the P.V. Engineering Forum (the name Piasecki Helicopter Corporation was not adopted until 1947), completed the first design, designated the PV–2 and using a 90-hp Franklin engine, ready for a first flight on 11th April 1943.

The United States Navy, anxious to find more powerful and capable helicopters for the demanding tasks awaiting the new aircraft at sea, was soon interested in Piasecki's ideas for a large twin tandem rotor helicopter, and on 1st February 1944, the USN awarded Piasecki a contract to design and develop his PV–3 tandem rotor helicopter. A successful first flight of the PV–3 prototype took place at the Piasecki factory at Manton, Pennsylvania, in March 1944, after which two further prototypes were built for USN evaluation, bearing the Navy designation XHRP–1. All went well, and the Navy evaluation was completed during the first six months of 1947, although as early as June 1946, the USN felt confident enough to order a pre-production batch of ten HRP–1 "Rescuer" helicopters for search and rescue and carrier plane-guard duties. The XHRP–1 used a single Wright R–975 piston engine, although production models were fitted with a tail-mounted 600-hp Pratt and Whitney R–1340–AN–1 engine.

The helicopter was growing up, fast. Shortly after the end of World War II, on 29th November 1945, the first rescue by the winch system, now so much a part of helicopter search and rescue operations, was made by a US Army R–5 helicopter. During a storm, a barge with a two-man crew aboard broke adrift and then ran aground on Penfield Reef in Long Island Sound, where it started to break up in the fury of the storm. The R–5, flown by Sikorsky's new Chief Test Pilot, Dmitry Viner, with US Army Captain Jackson Beighle manning the winch, had to make two trips since the small machine could only rescue one man at a time! However, the mission was accomplished successfully in winds of up to 60 mph. That same year of transition from war to peace, was also to see the first USN experiments with dunking sonar equipment using a Sikorsky R–6 helicopter, although this was still a far from practical proposition with the helicopter's stage of development and the limited performance of the day.

3
PEACE?

The helicopter emerged from World War II with its reputation firmly established, but it was a different world from the pre-war period during which the helicopter designers had developed their theories into workable designs. One of the lessons of the war had been the way in which the importance of strategic mobility had been brought home to the Allies, as the Japanese had quickly and successfully forced their way across the islands of the Pacific towards South-East Asia, and the Germans had invaded Crete in the first major airborne assault. The Allies themselves had learned many of the lessons, being able to mount an airborne assault on Arnhem, and then using airborne and amphibious forces together in the Normandy landings. If anyone had nurtured any hopes for the future of heavily defended entrenched positions on the battlefield, the collapse of the French Maginot Line in 1940 had dispelled them, for it had been a highly mobile war, with troops moving quickly over wide areas, using crack units to seize key positions in advance of the main force. The conflicts which were to follow were to continue this pattern.

Helicopters fitted into this scheme of things very well, but would fit even better if only they could be made larger and faster! In common with the jet aeroplane, the helicopter was a World War II concept which was to be forced to mature within a very short time due to the military, political and commercial pressures of the post-war world.

One of the main differences between the pre-war and post-war periods was the changed international situation. No thinking person could pretend that the world was at peace after the cessation of hostilities between the Allies and the Axis Powers, even if people had been able to forget about the problems of North Africa and China during the 1930s as both Italy and Japan pursued policies of colonization. The fall of Germany led very quickly to the installation of Communist puppet regimes throughout Eastern Europe, while Austria and Finland had neutrality forced upon them for their alleged misdemeanours during the war. The political position of Finland, which had been forced into an alliance with Germany during the war to resist Soviet aggression, became such after the war that the term "Finlandization" was coined to describe bogus nationhood, or suppression without occupation. Before long, the United States,

Canada, the United Kingdom, most of western Europe, Iceland, Greece and Turkey formed the North Atlantic Treaty Organization, NATO, into which West Germany was eventually admitted, although, for purely political reasons, Spain was left out until 1982, while Sweden, Switzerland and the Irish Republic clung to their traditional neutrality. NATO was a response to the Warsaw Pact, the enforced alliance of Russia's European satellites, from which only Yugoslavia and Albania managed to escape, although Romania was later to manage to remove Soviet forces from her territory while remaining a full Pact member.

Any attempt at rejection of Soviet control led to fast and firm intervention by Soviet forces, in Eastern Germany, in Hungary and Czechoslovakia, while the two alliances faced up to one another in Germany, northern Norway, Greece and Turkey, and on the world's oceans.

Meanwhile, the gradual collapse of the older European empires had been hastened by the stresses of the war, including the defeat of many colonial administrations by the Japanese in the Far East. The withdrawal of the older colonial powers was not always easy, and often involved fighting, sometimes in an attempt to hasten the day of departure, or sometimes in an attempt to sway the post-colonial administration of the territory. Not infrequently, Soviet or Chinese-backed Communist agents would attempt to create a state of chaos with the ultimate aim of ensuring that the newly independent power would subscribe to Marxist ideology. This type of conflict was to be one in which the helicopter would rule supreme, as the most valuable and adaptable of all of the different types of equipment available to the commander in the field.

At least one of the post-war changes was for the better. United States military aviation was entirely reorganized in 1947 with the formation, long overdue, of the United States Air Force, completely free of Army control, and coming just twenty-nine years after the formation of the Royal Air Force. This left the new USAF free to concentrate on developing America's strategic air power, while the United States Army could turn its attention to the proper development of organic air power, providing commanders in the field with the tactical air support required by ground forces. Further adjustments to the structure of United States service aviation followed, with the formation of a Military Air Transport Service, MATS, in 1949, to combine the separate air transport services of the individual air arms and remove some of the wasteful duplication of effort hitherto involved. This wise move still left some tactical air transport with the individual services, while MATS itself became a command within the USAF, and the armed forces retained their own helicopter units, linked to combat units, which over the years were to assume a growing combat role themselves. MATS was also to acquire helicopters later for the search and rescue role.

Helicopter development at the end of the war, saw Sikorsky R–4, R–5 and R–6 helicopters in service or entering service in the United States and Great Britain. Experiments continued, with Lieutenant Alan Bristow, RN, making the first helicopter landing on an escort vessel in September 1946, when he placed a Sikorsky R–4 Hoverfly on a British destroyer. Within a few months, in February 1947, Lieutenant K. Reed, RN, landed another R–4 Hoverfly on the newest British battleship, HMS *Vanguard*, setting down on the ship's quarterdeck.

Lieutenant K. Reed lands one of the Royal Navy's R–4s aboard the battleship, HMS *Vanguard*, early in 1947.

A development of the R–5 series with similar appearance was the S–51, seen here exercising with the US Marines. The nosewheel was a distinguishing feature.

A Westland-Sikorsky WS–51 Dragonfly helicopter, embarked aboard the light fleet carrier, HMS *Glory*, during the early 1950s.

Further records for the helicopter were being set at this time by one of the Sikorsky helicopters, the R–5A. On 14th November 1946, Majors Jensen and Dodds, US Army, established a duration record of 9 hours 57 minutes, and at the same time also established two other records, flying over a 1,000 kilometres closed circuit and a speed record for the distance of 107.25 kmph. Earlier that same year, Majors Cashman and Zins established a straight line distance record for the helicopter of 1,132.3 kilometres, again using an R–5A.

The excellent performance of the R–5 was wasted on the earlier versions, with their inconvenient narrow cabins and tandem seating for just two occupants, or even the three occupants of the later R–5D series. A further development took place, designated by Sikorsky as the S–51 in the company's new nomenclature for its helicopters, which remains in use today. S–51s provided four seats, and although produced with the civil market in mind, were soon being purchased by armed forces on both sides of the Atlantic, and beyond. After the first flight on 16th February 1946, followed by civil certification in March 1946, eleven aircraft were ordered for the USAF, designated R–5F in the original Army series, although a further order for thirty-nine S–51s with an up-rated specification for search and rescue duties was designated H–5G by the USAF, while another sixteen to the same specification but with floats instead of wheels were designated H–5H. In USN service, almost a hundred S–51s were designated as HO3S–1 and HO3S–2. Both USAF and USN S–51s were used for rescue duties, with the USAF aircraft operating within the newly-formed USAF Air Rescue Service with enclosed stretchers mounted on either side of the cabin, and as plane-guard and observation aircraft for the USN. A total of some 300 S–51s were built by Sikorsky between 1946 and 1951.

Meanwhile, Westland Aircraft, then a fairly small British manufacturer of fixed-wing aeroplanes, obtained a licence to construct the Sikorsky S–51 for the British market. Six US-built examples were supplied to Westland as part of this contract, and after the first flight of a British-built Westland-Sikorsky WS–51 on 5th October 1948, the British company produced another 138 of these machines, known to the British armed forces as the Dragonfly, until production ceased in 1953. The British WS–51s were fitted with an 520-hp Alvis Leonides engine, a type installed in a number of British aircraft, and although the early production HR Mk.1, of which thirteen were built for the Royal Navy, and HC Mk.2 for the Royal Air Force, of which three were built, were fitted with wood and fabric rotor blades, later versions, including fifty-eight HR Mk.3 for the Royal Navy and twelve HC Mk.4 for the Royal Air Force, used all-metal rotor blades.

This was an improvement, since there had been a number of accidents during the early days of the helicopter due to fabric-covered rotor blades ripping in flight, while during Arctic operations, the fabric had been known to harden and crack in the extreme cold. There can be little doubt that the fabric-covered rotor blades had been a source of weakness for the early helicopter.

Westland's relationship with Sikorsky was to be a long one. Under their agreement, the two companies exchanged information, and both companies developed from being relatively small manufacturers of fixed-wing aircraft to being major forces in rotary-wing aircraft production in North America and Europe; indeed, during the early 1950s, Westland was able to boast that the company was Europe's largest helicopter manufacturer, although this important distinction was allowed to slip in later years. Throughout the relationship, Westland retained the freedom to develop the original design, often to meet what the British company regarded as its own market requirements, and while the base market was in the United Kingdom, Westland-built examples of Sikorsky machines were also supplied to many countries in the British Commonwealth and to air forces and air arms throughout Western Europe and the Middle East. Non-British

A development of the WS–51 was the Westland Widgeon, seen here in civilian guise demonstrating its suitability for search and rescue duties.

The Bristol Sycamore was one of the earliest all-British designs, many of which were exported.

operators of the Westland Dragonfly, for example, included the Royal Ceylon Air Force, the Egyptian Air Force, Italian Air Force, French Air Force, Yugoslav Air Force, the Royal Iraqi Air Force, the Japanese Air Self-Defence Force and the Royal Thai Air Force. Many of these countries, including Italy, France and Japan, were later to reach their own licensing agreements with Sikorsky and other American helicopter manufacturers, or to develop designs of their own, accounting to some extent for the erosion of Westland's early position as the leading European helicopter manufacturer.

Later, in 1953, Westland developed the basic WS–51 into a new helicopter, the Westland Widgeon, utilizing the rotor head of the later Sikorsky S–55, and with a five-seat cabin similar in appearance to that of the Russian Mil Mi–1. The Widgeon first flew on 23rd April 1955, and although not used by the British armed forces, this machine was supplied to both civilian and military customers in Ceylon, Jordan, Brazil and Hong Kong.

Britain was at this time attempting to develop a truly British helicopter industry, with indigenous designs, appropriately enough for the country adopted by de la Cierva. Indeed, one of the early

British helicopter designs was a Cierva design, developed by G. and J. Weir, and known as the Air Horse, with a triple-rotor configuration to meet Air Ministry Specification E19/46. The Air Horse made its first flight in December 1948, with test pilot Alan Marsh taking the aircraft and three passengers to altitudes of between 500 feet and 1,000 feet with a load of ballast totalling another 1,250 lbs weight. However, this somewhat cumbersome machine never entered production.

An earlier British design, indeed the first British-designed helicopter, was the Bristol 171, designed by Raoul Hafner as the result of design studies commenced in June 1944. Two prototypes, with the designation 171 Mk.1, were built, and the first of these became the first British-designed helicopter to fly on 27th July 1947. A single-engined, single-rotor, four-seat design, the 171 was without a suitable British engine at first, and so an American 450-hp Pratt and Whitney Wasp Junior engine was used on the Mk.1. A single example of one Mk. 2, which first flew on 3rd September 1949, used a 550-hp Alvis Leonides 71 engine, as did the production Mk. 3, which was named the Sycamore, and had a wider cabin to allow a three-seat bench behind the pilot, making this a five-seat machine. This machine enjoyed some considerable success, with almost 200 Sycamores being built, initially for the British Army and for the Royal Air Force, with the Army machines being used for air observation post and communications duties, while the RAF machines were also used for communications duties, and for search and rescue, SAR, although before long, Sycamores were also being used on casualty evacuation duties in trouble spots in the Middle East and elsewhere. The first operational RAF Sycamore unit was No.275 Squadron, which received its first machines in April 1953, while the following year the type entered service with No.194 Squadron in Malaya, and in 1956, with No.284 Squadron in Cyprus.

The Sycamore did not enter service with the Royal Navy, but it was chosen by the Royal Australian Navy, which eventually operated ten Sycamores on communications, carrier plane guard and search and rescue duties, and by the Royal Australian Air Force, which received two, taking delivery of these at the same time as their Bristol 170 Freighters, some of which carried the Sycamores in their spacious cargo holds on their long ferry flights from Britain to Australia, possibly setting an early record for the longest flight as transport aircraft cargo by a helicopter! Certainly, loading the Sycamore into the 170, an aircraft originally conceived as a commercial car-carrying airliner, would have been a far easier task than loading the YR–4 into the Curtiss C–46. Other Sycamore customers included the Belgian Air

Force, which ordered three, and no less than fifty for the newly-formed West German armed forces, with most passing to the new German Army, the Bundeswehr, for liaison, communications and AOP duties, but both the Luftwaffe and the Bundesmarine or Navy, were to receive small quantities for SAR and communications duties before production ceased in 1959.

The Sycamore's capabilities were put to an early test by the British Army, in an exercise called 'Medusa', on Salisbury Plain, in 1952.

Rather less successful, in the commercial sense at least, was another Alvis Leonides-powered British helicopter, the Fairey Gyrodyne, although this single-rotor machine did establish a British rotary-wing aircraft speed record of 200 kmph, 124.3 mph, on 28th June 1948, after a first flight on 7th December 1947. The Gyrodyne was not a pure helicopter, but rather what would now be termed a compound helicopter, using a tractor propeller positioned in a fairing on the starboard wingtip, as well as a rotor. The original prototype crashed in April 1949, but a second prototype was completed and used for research purposes, although the Gyrodyne never entered production.

Another Cierva design achieved some modest success. Originally designated the Cierva W.14, this helicopter was designed to meet a British Army requirement for a small lightweight helicopter which could be used mainly for training purposes, although it would also have potential for AOP duties if needed. After a first flight on 8th October 1948, using a small 106-hp Jameson FF–1 engine, the Cierva W.14, by now also known as the Skeeter, started a prolonged development period, with a second machine, designated the Skeeter 2, using a 145-hp de Havilland Gipsy Major 10 engine and a modified tailplane of more conventional circular cross-section than the original triangular cross-section of the Skeeter 1 prototype. The Skeeter 2 flew for the first time on 15th October 1949. Early development was hampered by severe ground resonance problems, but work continued on the helicopter after Saunders-Roe acquired Cierva, and the design, in January 1951, to assist in competing for a Ministry of Supply contract. However, while further prototypes and pre-production machines were completed during the early 1950s using both 180-hp Blackburn Bombardier 702 and 183-hp de Havilland Gipsy Major 200 engines, it was not until 1958 that deliveries of the first of sixty-four production machines started. Most of the Skeeters built were used by the British Army at the Army Air Corps Training School, but a small number were also supplied to the Federal German Bundeswehr and Bundesmarine, again as training machines, although after a very short service life in Germany, these aircraft were passed to the Portuguese

Air Force in 1961.

The first of the United States Navy's Piasecki HRP–1 Rescuers, developed from the PV–3 prototype, made a successful maiden flight on 15th August 1946. A total of twenty production machines were built, making this the largest helicopter in service at the time, as well as being the first operational tandem-rotor machine. Nicknamed the 'flying banana', the Rescuer was a long, thin helicopter, with a fabric-covered forward fuselage and a distinctive angular fuselage shape which made its nickname singularly apt. The narrow fuselage cross-section meant that the crew of two had also to sit in tandem, although the extensive use of Plexiglass in the nose and sides of the flight deck or cockpit ensured an excellent view in most directions. Although seemingly long, the cabin could only accommodate eight passengers seated one behind the other in single seats, although it has to be recognized that even this somewhat limited performance marked a significant advance on all earlier designs, and that relatively good use was made of the single tail-mounted Pratt and Whitney R–1340–AN–1 piston engine rated at just 600 hp and driving two rotors. Power input for many of the smaller helicopters in production at the time – including some S–51 and WS–51 variants – was often not much less than that of the Rescuer, and their lifting capacity was frequently just half that of the Piasecki design. As an alternative to the eight passengers, the Rescuer could carry up to six stretcher cases, or one ton of cargo, which was usually slung under the fuselage.

Piasecki's first production helicopter, the HRP–1 Rescuer, was known as the 'flying banana', seen here aboard the USS *Palau* with one of the HRP developments in the middle.

Later, the Rescuer was to be used by the United States Marine Corps for assault training, and by the United States Coast Guard as a SAR helicopter, but in both cases, the machines used came from the original USN production and no additional examples were built. A development of the Rescuer, the HRP–2, designated the PV–17 by the manufacturer, used a larger and wider fuselage with side-by-side seating, all-metal construction, and had a number of technical innovations. It was built in limited numbers as an interim design paving the way for the later H–21 tandem-rotor helicopter, which was to enter production after Piasecki had changed its name in 1956 to Vertol, a company which still remains the world's leading manufacturer of twin-rotor helicopters.

Another form of twin-rotor helicopter, but of a less conventional layout, also appeared at this time, the Kaman K–125, designed by Charles Kaman. The K–125 made its first flight in January 1947, and used the novel feature of twin intermeshing or interconnecting rotors, a system which worked well on Kaman's early helicopters, but which tended to be at best baffling or at worst unnerving for spectators! The Kaman helicopters also differed from other designs by using a servo-surface on the rotor blades to alter the pitch and effect control. The K–125 was followed by the K–190 and the XHK–225, which made its first flight in April 1950, and which was also designated the HTK.

Another American to turn to the helicopter was Stanley Hiller, who, in 1944 at just eighteen years old, designed and built a helicopter called the XH–44, in fact the first twin-rotor American helicopter design to operate successfully. The XH–44 was rejected by its designer as being too heavy and costly to operate, and he then turned to single-rotor machines, developing a small helicopter known as the UH–5, which after further development became the Hiller 360, and eventually entered production as the Hiller Model 12 for civilian customers. The US armed forces soon became the major customer for this small helicopter, which became the H–23A Raven in US Army service, and the HTE–1 in US Navy service, in each case the helicopter was used mainly on training duties. The sixteen USN HTE–1s used a 200-hp Franklin 6V4–200–C33 engine, while a more powerful 210-hp Franklin unit was used in the US Army H–23As, of which one hundred entered service.

As one might have expected, given the penchant for helicopter design of the Russians, considerable effort was being put into helicopter design and development in the post-war Soviet Union; indeed, no less than three large and well-funded design teams were heavily committed to helicopter work. The first of these was that of

Nikolai Kamov, who had worked as an assistant to one of the other Soviet helicopter designers, Aleksander Yakovlev, but once in charge of his own design bureau concentrated on twin contra-rotating rotor machines for liaison and light transport duties. Kamov's first helicopter, the Ka–8, flew in 1947, but suffered through being severely under-powered, using just a 27-hp M–76 engine, and not surprisingly did not proceed beyond the prototype stage. Later, in 1949, Kamov built a more powerful helicopter, the Ka–10, which also flew, using just a 55-hp engine, but this was a single-seat machine of limited capabilities, although actually ordered into production in small numbers for the Soviet Navy to evaluate its potential in the fleet-spotting, or reconnaissance, role. Indeed, this was one of the first Soviet helicopters, if not the first, to be honoured with a NATO code-name, in this case 'Hot'. The first effective Kamov machine was the Ka–15, a two-seat machine using the Ka–10 rotor system but with a more realistic 255-hp Ivchenko AI–140 radial engine, which first flew in 1952, although it did not enter service with the Soviet armed forces, mainly the Navy, until 1955. The Ka–15 was used for reconnaissance, liaison and communications duties, and bore the NATO code-name 'Hen'.

All Soviet helicopters are given NATO code-names beginning with the letter 'H', just as bombers are given names beginning with 'B' and fighters' names beginning with 'F'. The idea is not to confuse or add some form of pseudo-dramatic effect, but to allow for the fact that some considerable period may elapse between the first sighting of a helicopter, or any other Soviet aircraft, and news of its official Soviet designation. The NATO designation is suffixed to distinguish between variants of the same machine. The system only appears to be strange to the layman in the West, who is used to reasonably comprehensive information about military projects from democratic governments and commercially minded manufacturers anxious for orders, and who are in sharp contrast to their Eastern counterparts.

Yakovlev himself had led one of the major Soviet helicopter design teams, but only for a short period. He designed and built a large tandem twin-rotor helicopter during the 1950s, but more of this in a later chapter. Yakovlev tended to take an intermittent interest in rotary-wing aircraft, and is better known for his work on fixed-wing utility aircraft.

The third and final member of this trio of Soviet helicopter design bureaux was Mikhail Leontyevich Mil, a college contemporary of Kamov, and who had also worked during World War II on the Bratukhin Omega, before being granted his own design bureau in

One of the most successful Russian helicopters was the first Mil design, the Mil Mi–1, which also was one of the first all-metal helicopters.

1947. Mil's first helicopter design, the Mi–1, appeared in September 1948. Unlike the early American designs, the Mi–1 used an all-metal fuselage instead of a metal tube framework fuselage covered by either fabric or plastic-impregnated plywood, but otherwise, the similarity to Sikorsky's work was close on this and on some later designs, with a single main rotor and auxiliary tail rotor driven by an engine mounted behind an enclosed cabin. Accommodation was provided for a pilot and up to three passengers, seated two abreast, while the power unit on production machines was a 575-hp Ivchenko AI–260 seven-cylinder radial engine. Details of the Mi–1 became known to NATO during 1950, as it began to enter service with all of the Soviet armed forces, but its designation and design bureau, with more complete information about capability, did not become known until the first recorded public appearance of the machine at the 1951 Tushino Air Display, by which time NATO itself had designated the aircraft the 'Hare'. Subsequently, the Mil Mi–1 was produced in vast quantities over a long production life spanning some two decades, and was built under licence in Poland by WSK–PZL, who also undertook some development of the machine, building an improved version with two small turbine engines known as the WSK–Swidnik Mi–2, or 'Hoplite' to NATO. A small quantity of Mi–1s would also seem to have been built, or at least assembled in Communist China.

Development after the end of World War II in the United States naturally covered a wide variety of helicopters, but none of these was more successful or remained in production longer than the Bell 47. This was a wartime design which entered large-scale production in 1946, with sixteen machines built for the United States Army to evaluate, while the newly formed USAF was to receive two 47s, with the pre-production designation YR–13, and the USN received ten broadly similar machines, designated the HTL–1. A small twin-seat cabin, with side-by-side seating, was fitted to the early military Bell 47A and the civil 47B, although an open cockpit version remained available for the hardy. The familiar 'goldfish bowl' moulded canopy soon appeared, during late 1947 on the Bell 47D, of which the US Army ordered sixty-five examples as the H–13B, while the US Navy ordered twelve as HTL–2s. A number of Army aircraft were converted to air ambulance versions as the H–13C, with stretcher positions on either side of the cabin, now able to accommodate up to three persons, and with the now familiar 'open' tailboom replacing the fabric-covered tailplane of the earlier versions. Production soon rose to what, at the

One of the longest production periods for any helicopter was enjoyed by the Bell 47, which first flew in 1946. This is a later 47G of the French Army.

time, seemed to be an astronomical figure for helicopters, with eighty-seven H–13Ds and almost five hundred three-seat, dual control H–13Es ordered by the US Army from 1949 onwards, while the United States Navy received forty-six of the equivalent HTL–4 and thirty-six three-seat, dual control HTL–5s. However, the definitive Bell 47 was to be the 47G, with the three-seat 'goldfish bowl' cabin of earlier versions, but with the original 200-hp Franklin engine replaced by a 200-hp Lycoming VO–435 engine after 1955, and which remained in production for more than twenty years after its introduction in 1953. The Bell 47G entered US Army service, with 265 aircraft designated the H–13G, and US Navy service, with forty-eight HTL–6s. This was the version of the 47 known as the Sioux in British Army service, and which was built under licence for the British Army by Westland and for other customers in Europe by Agusta in Italy. Lycoming-powered 47Gs were designated 47G–2 by the manufacturer, and almost five hundred of these went to the US Army as the H–13H.

Sikorsky also produced a small helicopter design during the late 1940s, the small twin-seat S–52, using a 178-hp Franklin engine, which made its first flight on 12th February 1947. The twin-seat version was designated HO5S in USN service, and a four-seat version, the S–52–2, with a 245-hp Franklin 0–425–1 engine entered US Marine Corps service with Squadron HMX–1 in March 1952, while the US Coast Guard received about eighty S–52s as HO5S–1Gs.

The uneasy peace of the immediate post-war period did not provide the helicopter with an early opportunity of proving itself on military or naval operations. The first of the many small wars to have plagued the Western democracies since 1945 arose as the French and the Dutch both strove to reassert their authority over their dominions in the Far East, lost to the Japanese as they had advanced during the early days of the war in the Pacific and South-East Asia. Nationalistic movements arose in these territories as the defeated Japanese departed, and in French Indo-China, the nationalists had backing from the Chinese Communists. Neither France nor the Netherlands was in a position to fight these colonial wars effectively, with their governments still struggling to repair the damage of wartime occupation and of successive battles on their homeland. Their armed forces were weary and ill-equipped to fight wars so far away from a home country which had known freedom for such a short time, and which many had not seen for five years of war while fighting with the 'free' forces alongside the other Allies. Helicopters could not play the significant role which they should have played in these campaigns in the Far East simply because only the British and Americans had such machines in any quantity, even though a few other countries, including Canada, were evaluating the helicopter. The French and the Dutch were, in any case, almost bankrupt.

Before many years had passed, both the French and the Dutch had abandoned their colonies in Asia, after fruitless and costly military campaigns. The Dutch were first to go, after encountering some strong international pressure to leave what was subsequently to become the Republic of Indonesia.

Britain also suffered problems in the Far East, with Communist bandits threatening the security of Malaya, including the rubber plantations on which so much of the region's prosperity depended.

One of the first Royal Air Force operational helicopter units, a flight of four Westland-Sikorsky Dragonfly HC2s, was dispatched to Singapore for casualty-evacuation duties in Malaya. Apart from the obvious military benefits of being able to fly troops to hospital in a very short time after they had been wounded, taking only hours for journeys which could have taken days or weeks through thick jungle or by meandering rivers, the availability of the helicopter to take sick civilians to hospital, or to fly in medical supplies or doctors and nurses, also played a major part in the campaign to win the hearts and minds of the local population, denying the terrorists the opportunity of gaining support in the remote rural areas. On their arrival in Singapore in June 1950, the RAF helicopters were based at Changi, and were soon being operated at full stretch. Possibly one of their most unusual passengers was the wife of a Communist bandit leader, found in the jungle only ten miles from the border with Thailand. A helicopter had to make a positioning flight from Singapore, flying 450 miles across thick jungle to the nearest British base, and leaving there on 30th November, with two pilots taking turns at the controls, flew the bandit's wife to Singapore, arriving there the following day. In this way, she was soon available for interrogation. On another occasion, two pilots flew for thirteen hours to take a badly wounded Army private to a base hospital.

Plans were also being made to provide a permanent patrol vessel for the British Governor of the Falkland Islands in the South Atlantic, which was also to be capable of carrying a platoon of Royal Marines and a helicopter for their rapid deployment ashore. The helicopter would have a mercy role as well, since the vessel would double as an ice patrol, undertake hydrographic survey work, and also help to support British scientists working in the Antarctic.

The first Royal Navy helicopter unit, No. 705 Squadron, had in fact been formed as early as May 1947, at Gosport, not far from the major naval base of Portsmouth, later introducing Westland-Sikorsky WS–51 Dragonfly helicopters, with a role as the pioneer squadron in developing techniques for the naval operation of helicopters, and to provide training for Royal Navy helicopter pilots, a role which was to remain with the squadron for many years. At first, this unit had operated the original Sikorsky helicopters, the R–4 Hoverfly I, and in contrast to some of the enthusiastic American descriptions of this machine, one British critic described it as a "girder-like contraption . . . covered in flimsy fabric with a glass house at the front end and a large propeller mounted horizontally on top, and which could carry little more than its crew while needing the full time of one man just to get it safely off the ground because it was so sensitive on the controls". That said, during the war years, the Royal Navy had already brought one of its early Hoverfly Is across the Atlantic aboard a ship, as a practical exercise to assess its value as a fleet submarine spotter, although as luck would have it, on this particular occasion, the helicopter was denied any worthwhile opportunity to prove itself!

Helicopters were already operating aboard American aircraft-carriers by this time. One USN pilot serving with a squadron embarked aboard the USS *Palau* was amongst the first to have good reason to be thankful for this new addition to the USN's aircraft inventory. On landing aboard the ship one day in March 1951, his Grumman Avenger swung out of control after its hook engaged the arrester wire, and the aircraft finished up hanging over the starboard side of the carrier's flight-deck, prevented from falling into the sea only by the tenuous hold of the arrester hook on the taut arrester wire. The arrival of the *Palau*'s plane-guard Sikorsky S–51, operated by USN Utility Squadron 2, enabled the Avenger pilot to be rescued from his cockpit without first falling into the sea, a hazardous operation so close to the side, and the screws, of a big ship.

In June 1951, when the Royal Navy submarine, HMS *Affray*, failed to report while on a training dive off the southern tip of the Isle of Wight, the full emergency procedures were put in hand, while the Flag Officer, Submarines, was able to fly over the search area in a Westland-Sikorsky WS–51 Dragonfly. Unfortunately, the submarine had been lost with its crew, and in this instance the helicopter had no opportunity to demonstrate its rescue capabilities.

Meanwhile, Sikorsky was working on a larger helicopter, one which could rival helicopters of the size of the Piasecki HRP–1 Rescuer, while retaining the single-rotor layout to which the company was now finally linked. This helicopter, designated by the manufacturer as the S–55, was to outstrip the entire production of the combined R–4, R–5,

61

and R–6 series. One significant difference between the S–55 and the earlier helicopters was the positioning of the engine in front of the passenger cabin, in the nose, while the cockpit for the two-man crew was moved above the cabin and in front of the rotor transmission. First flight by a prototype, designated YH–19 by the United States Air Force, was on 10th November 1949. Five pre-production YH–19s were built, the forerunners of some 1,300 S–55 helicopters to be built for civilian and military customers by Sikorsky, while the British licensee, Westland, was to build more than 400 S–55s, marketing these under the name of the Westland Whirlwind, with both Mitsubishi in Japan and Sud-Est in France also building the S–55 under licence, although in far smaller quantities than in the United States or Britain.

The initial production of the S–55 for military customers included fifty search and rescue H–19As for the United States Air Force, seventy-two H–19C Chickasaws for the United States Army, and ten HO4S–1s for the United States Navy. The first delivery of the naval version was to utility squadron HU–2 on 27th December 1950, initially for use as a general-purpose helicopter and anti-submarine spotter, but this was the helicopter which was to play the leading role in the further development and refinement of helicopter-operating techniques. The US Army Transportation Corps received their first H–19C Chickasaws in January 1952, which, with their ten-passenger capacity, proved a significant improvement in helicopter-borne

Designed to compete with the larger Piasecki helicopters, the Sikorsky S–55 saw considerable service with the USAF as the H–19A, used primarily for search and rescue.

mobility. Early machines used a 550-hp Pratt and Whitney R–1340 engine, but the later USN HO4S–3s, of which sixty-one were built, used the 700-hp Wright R–1300, while the United States Coast Guard ordered thirty HO4S–3Gs, although these were redesignated HH–19Gs in 1962.

In terms of capacity, the S–55 was a substantial improvement, even over the HRP–1. The ten passengers who could be accommodated inside the cabin sat with three with their backs to the engine, facing rearward, three at the back of the cabin facing forward, while on each side, there were two seats facing inwards. Large doors helped in unloading troops quickly in a forward combat zone, or in rescue operations. Folding rotor blades simplified shipboard stowage, including striking down into the hangars of aircraft-carriers. The position of the engine also made changing an engine a relatively simple exercise, assisting in recovering helicopters which had made forced landings in remote locations.

One of the early distinctions which fell to the S–55 was the first transatlantic helicopter flight by two Sikorsky S-55s, or H-19As, of the Military Air Transport Service Air Rescue Wing, positioning from East Hartford, Connecticut to the USAF base at Wiesbaden in West Germany. The two helicopters were flown by Captain Vincent McGovern and Lieutenant Harold Moore, USAF, with Captains Jeffers and Gombrich as their co-pilots. Modifications were made to increase the fuel capacity of the two S–55s from the standard 180 gallons to 480 US gallons, and after leaving East Hartford on 14th July 1952, they arrived, via Greenland and Iceland, at Prestwick in Scotland sixteen days later, on 31st July, having taken 45 hours 45 minutes flying time to cross the Atlantic. The total journey from the United States to West Germany took twenty days and fifty-two flying hours per machine.

It must have been a tedious way of crossing the Atlantic, with a normal cruising speed of just over 90 mph, and a maximum speed of 112 mph, while the extra fuel tankage was vital, with a normal range of just over 300 miles!

At the opposite extreme, the small American firm of Hoppi-Copter Inc. of Seattle, a town famous for very large Boeing aircraft, was working on the development of a small 'strap-on' helicopter to meet a special military need for individual helicopters for infantrymen, providing them with fast battlefield mobility. An early helicopter of this kind had been demonstrated in 1945, and taken to England for further demonstrations in 1948, although that same year a further development featured a built-in undercarriage, as an alternative to the legs of the infantryman! The manufacturer reverted to the original strap-on concept in 1952, but the idea did not advance much beyond this. The obvious appeal of large numbers of infantrymen advancing across a battlefield, each with his own helicopter, has to be set against the limited capabilities of such machines, unable to fly very fast or very far, the loss of the ability to carry very heavy loads, which is one benefit of large troop-carrying helicopters, and the appalling cost in terms of equipment and training for such a mass of helicopter-borne infantrymen, doubtless all of whom would expect pilot's flying pay!

Meanwhile, war threatened yet again. International tension had been rising for some time, in Europe and elsewhere, as the former World War II Allies broke up in the face of overt territorial ambitions by the Soviet Union, which had by this time assumed control of much of Europe and had already turned its attentions to Asia, assisted by the Communists' take-over of China. In spite of agreement over the post-war partition of Germany, and of the former capital, Berlin, and of British, American and French access to Berlin, the Soviet Union had attempted to blockade the city, and failed, thanks to a massive airlift by the Allied powers, in which the helicopter, at its then limited stage of development could not play a part. It was, on the other hand, one of the last major operations for the flying boat, ultimately to be one of the first main victims to the advance of the helicopter, along with the convoy-protection airship and the assault glider.

On 25th June 1950, without prior ultimatum or declaration of war, North Korea invaded the Republic of South Korea, launching a well-co-ordinated, full-scale attack both on the ground and in the air. The United Nations Security Council demanded an immediate cease-fire and withdrawal, and for possibly the first and last time in its history, was able to take firm and decisive action due to a Soviet boycott of the Security Council at the time. Normally, the Soviet Union, one of the permanent Security Council members, can veto any move contrary to its interests or that of its satellites. A United Nations force was hastily assembled and dispatched to Korea, taking rather longer than might have been hoped for, partly due to the still war-battered condition of Europe, and to the preoccupation of both the United States and Great Britain with policing and controlling occupied Germany and Japan. However, an advance force of naval vessels was quickly assembled by the main Western allies, with the new British light fleet carrier, HMS *Triumph*, arriving in Korean waters within five days of the outbreak of hostilities, along with the cruisers HMS *Belfast* and *Jamaica*, and five escort vessels, joining the ships of the United States Seventh Fleet.

Although the Royal Navy was equipped with helicopters by this

The Korean War saw the helicopter used in action, with many units using Sikorsky S–51s, such as this one, for communications duties.

time, *Triumph* did not have a helicopter of her own to act as plane guard for her carrier air group, and so a United States Navy Sikorsky S–51 helicopter, with a pilot, was loaned to the British ship and remained with her until she left Korean waters to return to Britain on 9th October 1950. The successor to *Triumph* was HMS *Theseus*, a sister ship of the same class, and she also enjoyed the support of a United States Navy Sikorsky S–51 during flying operations throughout the appalling Korean winter, of such severity that it reminded many Royal Navy men of the severe conditions encountered on wartime convoys to Russia.

Plane guard was an invaluable service, particularly so in the days before the angled flight-deck allowed aircraft to overshoot and fly round for another attempt at landing on. However, this important duty was but the tip of the iceberg as it were for the helicopter, which throughout the Korean War was to be used by the Americans for a wide variety of duties over a difficult and mountainous battlefront. This was a fast-moving war, and at one time the rapid Communist advance seemed to be unstoppable, but the Allied counter-attack was no less mobile, allowing little for the difficult terrain which would normally slow ground forces. Fortunately, both the United States Navy and Army were already pushing their respective versions of the S–55 into service in quantity, while Westland started Whirlwind deliveries on the first anniversary of the outbreak of war.

The Korean War saw the first mass transport of troops to the battlefront by helicopter. On 20th September 1951, twelve Sikorsky S–55s of the United States Navy lifted a company of 228 fully equipped US Marines to the top of a strategically important 3,000 foot high hilltop in central Korea, following this initial airlift with the delivery of nine tons of food and other supplies, before completing the operation by laying telegraph wires back to the area headquarters. Without the helicopters, the troops would have taken two days to move onto the mountain, and would have been exposed to enemy attack and the risk of losing their objective throughout this period, with heavy losses being inevitable, but by helicopter, the entire operation took just four hours.

Within a month, the same helicopters were back in action, moving a battalion force of 1,000 combat-equipped US Marines to the front line in east central Korea, in full view of enemy forces. Again, the operation took but a fraction of the time which surface movement would have required, just 6 hours 15 minutes, which was 25 minutes less than the Marines had planned! By the end of January 1951, every divisional general in Korea had at least one helicopter at his disposal for liaison and light transport duties.

These successes were the results of some four years of active research and exercising by the United States Marines, who had used twelve of the USN's small fleet of Piasecki HRP Rescuers to practise airborne, or helicopter-borne, assaults. Strictly speaking, the Korean operations were not assaults, but battlefield transport operations, but the basic requirements of well-drilled movement by the troops to ensure fast loading and unloading, and the flying techniques, differed little. A new phrase was coined to describe the rapid movement and subsequent deployment of troops by helicopter, 'vertical envelopment', emphasizing the ability to overcome obstacles and surround enemy forces. Surprisingly, in spite of its low speed and the vulnerability of its rotors to enemy attack, the helicopter did not show itself to be unduly at risk during the Korean War.

One exercise held during the Korean War, at Quantico, Virginia, in 1952, used United States Marine Corps troops aboard the USS *Palau* to investigate the possibility of moving troops in force by helicopter from ships off an enemy coast to key positions inland, bypassing the beaches on which enemy defensive positions would be based, and which might also be mined! Again, the HRP Rescuer was the helicopter used in this successful exercise.

Another notable exercise used the HRS–1 version of the S–55 for assault and battlefield transport evaluation of the new procedure known to the US Marines as 'hit an' git'. This meant a system of flying in a rocket launcher with its crew to a suitable firing position, landing, firing a few rounds at the enemy position, then moving on by helicopter taking the still hot rocket launcher to a fresh firing position before enemy artillery could locate the rocket firing position. The rocket launcher was slung under the helicopter while in transit. Fine though this appeared to be in theory, and during exercises, it was seldom used in practice.

No less impressive were the statistics. Compared with World War II, deaths amongst casualties reaching first-aid posts in Korea were down from 4.5 per cent to just 2 per cent, simply as a result of being able to use helicopter casualty evacuation, or CASEVAC. The extensive use of the helicopter in this role during the Korean War shortened the time necessary to get a wounded man to a first-aid station, and reduced the very real risk of further injury or complications arising during transport over rough ground and indifferent battle-front roads.

Obviously the wounds and injuries requiring the most careful handling and least suited to rides in a bumpy jeep or field ambulance were the head and stomach wounds. Before the Korean War, or perhaps one should really say, before the arrival of the helicopter, between 80 and 90 per cent of soldiers with such wounds died, but fast and relatively smooth helicopter transit to hospital reduced this appalling figure to 10 per cent. The S–55, in both its H–19 and HRS forms, was operating CASEVAC services in Korea from April 1951, onwards, along with the Bell 47.

The problem of communications in difficult terrain had helped the aeroplane to become more easily and widely accepted in areas where high mountains or dense jungle made surface communications slow, difficult, uncomfortable and expensive, but the one proviso had to be the availability of a landing site of suitable size. One answer to this problem lay in the flying boat or amphibian, but suitable stretches of water were not always available, and anyway, there were, and are, still some additional hazards in a water landing, including the lack of navigational aids, the deceptive effect of sunlight on water, underwater obstructions, and the problems of water which is too rough, or too smooth, in the latter case making "unsticking" during take-off difficult. The helicopter showed at an early date that it was the answer. Indeed, in 1955, one 1944 vintage R–5 helicopter of the Military Air Transport Service's 4th Rescue Squadron, helped to quell an outbreak of yellow fever in Costa Rica, flying medical teams to innoculate a

(*Above left*) The Korean War saw the helicopter prove its worth on CASEVAC duties, with the Bell 47 playing a prominent part. The stretcher's own Perspex canopy can be seen clearly in this picture.

(*Above right*) Another Korean War shot of a Bell 47 busy on CASEVAC duties.

thousand persons in the inaccessible regions of north and central Costa Rica.

Other armed forces were receiving the helicopter by this time. The Canadian Army obtained a Sikorsky R–6A Hoverfly II soon after the end of World War II with which to conduct its own trials. Before long, two air observation flights were operating helicopters, one with Bell H–13s, or 47s, and the other with Sikorsky H–19s, or S–55s. The formation of the North Atlantic Treaty Organization was accompanied by a large US-supported Mutual Defence Aid Programme, which included supplies of Sikorsky and Bell helicopters for America's allies, mainly in Europe. Many Latin American countries received military aid under the auspices of the Organization of American States. One of the earlier Latin American beneficiaries of American generosity was Brazil, whose armed forces, received twelve Bell HTL–3 (Bell 47) helicopters during 1950 and 1951 under the Military Aid Programme.

A characteristic of the Korean conflict, and of many of the other small and often dishonourable conflicts which have followed, is that while negotiations on a cease-fire continued, so too did the fighting, and it was not until July 1953, that the Korean War ended. The helicopter had proved itself to be invaluable, although only used on the Allied side, and many lessons had been learnt, which could only be fully understood and appreciated under actual combat conditions. Some of the US helicopter designs, including the Hiller H–23, were modified as a result of the lessons learned in Korea, but, as usual, the main result was pressure for larger and faster helicopters, in greater quantity. This was the main effect of the Korean War on the helicopter, but it should also be remembered that while the helicopter was working in Korea, other helicopters had been taking part in other conflicts, such as the Malayan emergency, which were to be no less typical of its future role. There was another result of the Korean War. The consequent rift between the East and West forced the Allies into the realization that they could no longer afford to tie up large forces policing occupied Japan and Germany. Consequently, first rearmament of Japan and then of Germany followed, with the Japanese creating first a para-military police force before developing this into the so-called self-defence forces.

4
AT SEA

As we have already seen, the navies of the world did not need a great deal of persuading of the value of the helicopter. The limitations of the landplane and the sheer impossibility of ever being able to afford the cost or the manpower for sufficient aircraft-carriers to adequately protect numerous wartime convoys, sometimes quite small, operating across wide oceans in all parts of the world, meant that naval commanders could readily appreciate the true worth of a machine which could operate from a small platform, possibly even aboard merchant vessels. Although airships were used with considerable success by the United States Navy during World War II on convoy escort duties, their lack of speed and the poor manoeuvrability of these unwieldy craft were obvious limitations, even though no ships were lost to enemy submarine activity from an airship-escorted convoy.

Even when aircraft-carriers were available, it was obvious that the helicopter was the right answer to the problem of plane-guard duties, essential because of the peculiar operational risks of naval flying, but which had previously required a destroyer to wait in attendance on the aircraft-carrier. It was soon clear that the destroyer could not reach the crew of a downed aircraft as quickly as a helicopter, and of course the destroyer was at risk while stopping in a combat zone to pick up survivors from an aircraft which had overshot the carrier's flight-deck. Against the two- or three-man crew of a rescue helicopter, a 1940s or 1950s destroyer usually had a crew in excess of 300 men, yet smaller vessels than a destroyer lacked the speed and sea-keeping capability necessary to maintain station on a large and fast aircraft-carrier in the open ocean.

In naval aviation, one also has to include the United States Coast Guard, which during time of war operates under the control of the United States Navy rather than the Treasury Department (it has since been passed to the Department of Transportation). So impressed was the United States Coast Guard by the results of early trials with the helicopter, including the Sikorsky R–4, that, as mentioned earlier, they established one of the first helicopter training schools at Floyd Bennett Field in November 1943, while continuing to participate in the trials with the United States Navy and the Royal Navy. The Coast Guard involvement also extended to manning the uss *Governor Cobb*, the American trials ship.

The rate of development of the helicopter and its operational acceptance by the navies involved can perhaps be best illustrated by the fact that, within a year of its opening, the helicopter training school at Floyd Bennett had trained more than 100 pilots and a large number of mechanics, and, before long, large numbers of Royal Navy air- and ground-crew were also to be trained here before the British could devote the resources to establishing their own training facilities.

Amongst the early triumphs of the USCG's helicopter operation was the rescue of the crew of a Royal Canadian Air Force aircraft which had crashed in the frozen wastes of Labrador on 19th April 1945. Scout planes discovered the stranded Canadians, but when ski-planes were sent to the rescue, one crashed attempting to land on a lake and another was stranded in thawing snow after taking on board two survivors. A Coast Guard Sikorsky R–4 from Floyd Bennett was hastily dismantled and flown in a Douglas C–54 cargo plane to Goose Bay, where it was reassembled, and then ferried 150 miles to a forward base from whence it conducted the final rescue operations at a radius of 32 miles from the forward base, picking up survivors from a hillside some 2,000 feet above sea-level in freezing weather conditions. Each round trip took about 90 minutes. Later, when a Belgian airliner crashed on the approach to the airport at Gander, Newfoundland, in 1946, only a helicopter could rescue the eighteen survivors, taking them to a flying boat for the flight to hospital. This was, of course, in addition to the mercy flight after the explosion aboard the destroyer USS *Turner* in January 1944.

Nevertheless, it was soon also apparent that there were some severe limitations to the possibilities for early naval helicopter operation. When, on 2nd January 1944, the British Helicopter Service Trials Unit embarked on the SS *Dagheston*, taking two helicopters, a YR–4B and an R–4, to ascertain their suitability for operations at sea, with five Royal Navy officers, four from the United States Navy and one from the Coast Guard, they found themselves aboard a ship prone to excessive rolling and pitching due to her heavy loading. Not once, according to USCG reports, did the ship roll through less than twenty degrees, that is ten degrees to port and to starboard, and at times the incidence of roll increased to forty-five degrees to port and to starboard. A flight-deck or landing platform had been installed aboard the ship, some 96 feet long and 50 feet wide, 22 feet above the water-line and with a 17-feet high windscreen along the forward edge of the platform, but operations were only possible for three of the sixteen days during which the ship was at sea, and on two of the three days,

71

flying was reduced to just 30 minutes duration! Although the R–4 suffered an 80-mph gale without damage to the rotor blades, the entire operation was hazardous and it was concluded at that time that twenty degrees of roll was the maximum which could be sustained during naval operations of the helicopter, while the R–4 lacked the power to cope with the effects of heavy pitching, being unable to follow the heaving deck. Naturally, these problems were in addition to the R–4's inability to carry anti-submarine weapons.

However, tests on anti-submarine operations followed, using anti-submarine underwater sound-detection equipment, which to the surprise and delight of all concerned was found not to be affected by the downbeat from the helicopter's rotors. This still left the problem of the helicopter's limited warload unresolved for the time being.

Experiments on rescue winches and harnesses took place during the autumn of 1944, with tests on picking up survivors from the water and from a small boat travelling at speeds of up to 20 knots. On 3rd October, at Manasquan, New Jersey, one USCG helicopter demonstrated its ability to rescue four men, one at a time, picking these up from life rafts and transferring them to the *Governor Cobb*, taking just ten minutes to rescue all four. As the trials progressed, early in 1945 a harness was developed which could be dropped to a person in the water, followed by another harness which could be thrown over a person in the water using a fending pole, so ending the need for crew members to actually get into the water during a rescue attempt; but although successful it never found widespread acceptance.

Of course, as mentioned in an earlier chapter, the first rescue by a helicopter using a winch was during late 1945, with a helicopter from the Sikorsky factory taking two men off a barge, which had run aground in a storm.

Helicopter pilots themselves sometimes ran into difficulties. The ability of the helicopter to autorotate more or less safely towards the ground after an engine failure minimized the effects of early engine failures, but bailing out of a rotary-wing machine has never been easy, due to the downwash of the rotor blades and the awesome prospect that, in anything more serious than an engine failure, control was lost immediately. The distinction of being the first man successfully to bail out of a helicopter belongs to Master Chief Aviation Mechanics Mate John Greathouse, of the United States Coast Guard. Based at Floyd Bennett Field, Greathouse and his mechanic, John Smith, were ordered to fly an HOS helicopter to Philadelphia to assist in the calibration of a warship's anti-aircraft gun radar on 25th September 1945. Unusually, the seat cushions of the helicopter were removed and the two men took parachutes with them. Flying in gusty conditions over Philadelphia and the shipping in the bay, Greathouse and Smith heard a loud bang followed by loss of control, after which they quickly bailed out through the helicopter's large side windows. Both men landed safely, after the strong winds blew them into the centre of the town. It seems that a rotor-blade assembly had failed, leading to the loss of control, and in fact only the opportunity to bail out presented by the availability of their parachutes saved their lives. Bailing out in this way from a helicopter is, in fact, a rarity and few helicopter pilots carry parachutes.

Greathouse, incidentally, on his retirement from the USCG more than twenty-five years later, was the last serving non-commissioned pilot in the service, although his last years as an aviator were spent flying fixed-wing Grumman Albatross amphibians!

Meanwhile, on the other side of the Atlantic, the Royal Navy was soon ready to move on from the R–4 helicopter and the limitations of its performance. In May 1947, the first British naval helicopter squadron, No. 705, was formed at Gosport to handle helicopter trials and continue the service evaluation of helicopter-operating techniques, using Sikorsky R–4 Hoverfly I machines. After further trials and experiments, the first of the 139 Westland-Sikorsky WS–51 Dragonflies to be built by Westland was delivered to the Royal Navy early in 1949, and joined the light fleet carrier, HMS *Vengeance* for trials in the Arctic Ocean, intended to test a variety of new naval aircraft and equipment, with the helicopter performing satisfactorily in this demanding environment. Later that same year, No. 705 Squadron received its WS–51 Dragonflies, and the Royal Navy started to put at least one plane-guard machine aboard each of its aircraft-carriers, although this was not finally accomplished until just after the end of the Korean War.

A further series of trials with the WS–51 took place aboard the British Royal Fleet Auxiliary, *Fort Dusquesne*, an 11-knot wartime 'Victory' ship, in the English Channel, to evaluate further the potential of the helicopter as an anti-submarine weapon operating from merchant vessels.

The WS–51 gradually replaced the earlier helicopters in both Royal Navy and Royal Air Force service, and its vastly improved performance helped the helicopter's progress from the experimental role to the operational one. However, further improvements were still to follow. In 1950, Westland proudly announced that it had been awarded a licence to build the Sikorsky S–55 in the United Kingdom, and before long both the Royal Navy and the Royal Air Force ordered

The Westland Whirlwind was a popular British-built version of the S–55; seen here in RAF service.

this classic helicopter design. However, before the first Westland-built S–55s, or Whirlwinds, could be delivered, the first S–55s for the Royal Navy were delivered from the United States in 1952 under the Mutual Defence Aid Programme. These were standard USN machines with Wright Cyclone piston engines, a 110-mph top speed and a range of almost 400 miles, and were delivered to the second Royal Navy helicopter squadron, No. 848.

After a few months spent 'working up' in the UK, No. 848 Squadron and its Sikorsky S–55 helicopters were sent to Singapore, sailing aboard the light fleet carrier, HMS *Perseus*, to help the hard-pressed Royal Air Force S–51 flight which had been operating so valiantly for some two years in the jungles of Malaya. No. 848's S–55s arrived in Singapore on 8th June 1953, and were flown ashore. There was a slight hitch; one of the helicopters suffered an engine failure on the flight from the carrier to the base at Changi, and made a forced landing without injury to the crew. The machine itself was also subsequently recovered after an engineer was winched down to rectify the fault, demonstrating the serviceability of the S–55 design. During their first year in Malaya and Singapore, the helicopters of No. 848 Squadron carried 200,000 lbs of freight, and some 10,000 troops, handled a large number of casualty-evacuation cases, landed paratroops and tracker dogs, dropped leaflets and conducted reconnaissance flights, spending some 3,500 hours in the air. Later, the squadron returned home to be equipped with Westland Whirlwind helicopters with dunking sonar for the anti-submarine role.

The first operational helicopter to be fitted with the dunking sonar, the S–55, as already mentioned, suffered from power limitations which meant that these helicopters could not operate in the self-contained hunter-killer role, but instead were divided into sonar-equipped hunters and torpedo or depth-charge-armed killers. The Royal Navy also received its first anti-submarine S–55s under the MDAP programme, with the delivery of fifteen Sikorsky S–55 HO4S–3 helicopters as the Fleet Air Arm's HAS Mk. 22 to the then new No. 845 Squadron based at the Royal Naval Air Station, Hal Far, Malta, during the mid-1950s. These machines, the later British-built but American-powered HAR Mk. 4, and the 750-hp Alvis Leonides HAR Mk. 5, were all used to replace the Fairey Gannet carrier-borne anti-submarine aircraft which had first entered Fleet Air Arm service just a few years earlier, in 1953, since the helicopter with its dunking sonar was felt to be a more useful anti-submarine detector than a fixed-wing aeroplane.

The Royal Navy also received twenty Hiller HTE–1 light helicopters for pilot training, replacing the remaining Sikorsky R–4s and Westland WS–51 Dragonflies at Culdrose, in Cornwall, which was also the base of the training squadron, No. 705 again, for many years. In Fleet Air Arm use, the HTE–1 became known as the H–1.

The Whirlwind and its American counterpart, the S–55, were entirely successful helicopters, able to handle a wide variety of roles, including movement of sufficient numbers of troops or casualties to be

By the mid-1950s, carrier operations without a plane-guard helicopter were almost unthinkable, although these anti-submarine Fairey Gannets were soon ousted by ASW helicopters.

The need for a helicopter able to combine the ASW hunter and killer roles resulted in the Bell HSL–1, shown here dunking its sonar. The machine did not enjoy great success.

This ground shot of the HSL–1 clearly shows the combination of the Bell rotor system with tandem-rotor layout.

regarded not simply as a worthwhile addition to the inventory of any armed service, but indeed as indispensable. Nevertheless, as the helicopter developed, the idea occurred that greater specialization should be designed into the next generation, and the first helicopter to receive this treatment was an anti-submarine helicopter for the United States Navy, the Bell Model 61, designated by its sole customer as the HSL–1, the first and only tandem-rotor helicopter from this manufacturer. The traditional Bell two-bladed rotors were retained for the HSL–1, which also used two tail-mounted 400-hp Pratt and Whitney R–2800–80 piston engines to provide a combined hunter-killer machine for the first time, fitted with dunking sonar and the ability to carry anti-submarine depth-charges or torpedoes, or the then new Fairchild Petrel air-to-surface anti-shipping missile. Bell was awarded the contract to build the HSL–1 after winning a design competition in June 1950, and after building three prototype XHSL–1 helicopters, the company received a production contract for seventy-eight machines, of which eighteen were intended for the Royal Navy's Fleet Air Arm. Deliveries started in June 1957 to USN Squadron HU–1, but after some fifty machines had been delivered, it was decided to end production, with the HSL–1 remaining in service for only a short period, while the Royal Navy never received those HSL–1 helicopters originally intended for it.

Kaman also built specialized helicopters for the United States Navy, with a development of the original HTK–1, known as the K–600 to the manufacturer, and designated HUK–1 in USN service and HOK–1 in USMC service, and retaining the manufacturers distinctive intermeshing rotor system. The HOK–1 used a 600-hp Pratt and Whitney R–1340–48 while the HUK–1 used a Pratt and Whitney R–1340–52 engine, but of similar power output. Ordered in 1950, after successful prototype trials, eighty-one machines were built for the USMC and twenty-four for the USN, being used primarily for plane-guard and rescue duties. Later, the Kaman K–600 was also used as the USAF's Huskie, the H–43, later HH–43B, rescue helicopter, performing this role, and those for the USN and USMC, with considerable success.

A much larger helicopter than the eight-passenger Huskie or even the Bell HSL, was the Vertol HUP–1 Retriever, originally started by Piasecki before the change of name to Vertol in 1956, as the prototype PV–14 or, in United States Navy designation, CHJP, of which two were built to assess their suitability for a US Navy requirement for a utility helicopter to be used on a wide variety of operations, including carrier plane-guard, search and rescue and transport or utility

77

The ultimate development of Charles Kaman's series of intermeshing rotor helicopters was the SAR HH–43B Huskie of the USAF.

An early USN example of the Piasecki HUP–1 Retriever, note the tail-fins which were absent from later versions.

operations. After careful evaluation of the prototypes by the USN, thirty-two production PV–18s or HUP–1 Retrievers, were built, powered by a single 525-hp Continental R–975–34 piston engine and capable of carrying up to five passengers or three stretchers as well as a two-man crew. Folding rotor blades assisted in striking down into the hangar of an aircraft-carrier. However, the main feature of the Retriever was the provision of a large hatch in the floor of the fuselage, offset to starboard, through which a winch capable of lifting up to 400 lbs could operate.

Deliveries of the HUP–1 started in February 1949, by which time work was proceeding on the HUP–2, which used a Sperry autopilot and lacked the tail-fins of the earlier machine, as well as having an up-rated Continental R–975–42 engine. The first of the 165 USN HUP–2s was delivered in January 1951, followed by fifteen for the French Aeronavale and a small number with dunking sonar, designated the HUP–2S. In 1951, the United States Army ordered seventy HU–25A Mules, as the military version was known, and these entered service in 1953, and a number of this version was also built for the USN, as the HUP–3, and for the Royal Canadian Navy, for use on CASEVAC and light transport duties.

During its period in USN service, it was the HUP–2 which was used for trials with watertight hulls, probably the first experiments of their kind.

Not all large helicopters required twin rotors, indeed, the number of rotors has tended to depend more on the design philosophy of the designer or manufacturer than on practical considerations of size, although the complication of a twin-rotor design is obviously wasted on a small helicopter while the additional lift of twin rotors is equally an asset for larger helicopters. Sikorsky had remained faithful to the

(*Bottom left*) The bulk of the Retriever production consisted of the HUP–2 version, shown here.

(*Bottom right*) In US Army service, the Retriever became the HU–25A Mule.

single-rotor concept despite increasing helicopter sizes. The first significant attempt by Sikorsky to build a large transport helicopter came with the S–56, designed to meet a United States Marine Corps requirement in 1950, and which for almost a decade after its first flight, as the XHR2S–1, on 18th December 1953, was still the largest helicopter operating outside the Soviet Union. Altogether no less than fifty-five examples of this big helicopter were delivered to the United States Marine Corps, starting with Squadron HMX–1 on 26th July 1956. One of the first single-rotor helicopters to use twin engines, the two 1900-hp Pratt and Witney R–2800–54 engines were mounted on outriggers to leave the entire space below the cockpit and the engine gearbox free for cargo or other loads. This effect was enhanced by the use of large clam-shell nose doors and a 2,000-lb capacity electric winch at the back of the fuselage to assist in loading heavy equipment which, as an alternative to the maximum capacity of twenty-four troops or twenty stretchers, could include an assortment of small vehicles or artillery pieces. A rear fuselage side door was also fitted.

One interesting development of the HR2S–1 was the HR2S–1W, an airborne early warning, AEW, variant of which two were flown in 1957, with a large downward-looking radar scanner for an AN/APS–20E radar mounted in the nose, giving the helicopter a distinctive heavy chin and also demanding modifications to the tailplane in order to maintain the trim of the aircraft. Big though it was, the S–56 series was also fast for its day, with a 130 mph top speed, a cruising speed of 115 mph and a range of 145 statute miles, a predecessor in many ways of Sikorsky's brilliant S–60 series. Most of the contemporaries of the S–56 were operating at maximum speeds little in excess of 100 mph, and with cruising speeds still often below the magical 100-mph figure, although most could show a better range than the S–56.

Not all of the helicopter developments at this time belonged to the American manufacturers. The British Bristol Sycamore Mk. 50 entered service with the Royal Australian Navy in January 1953, initially on carrier plane-guard and communications duties aboard HMS *Vengeance*, the light fleet carrier loaned from the Royal Navy while the Australians awaited their own HMAS *Melbourne*. Conventional search and rescue duties also fell to the small Sycamore, with more than seventy lives being saved by *Vengeance*'s three Sycamores after severe flooding in New South Wales, again during 1953. Altogether, the RAN's Sycamores saved twenty-two lives and rescued a total of 200 others from less serious predicaments in search and rescue and disaster relief during the five years after they had replaced the loaned United

States Navy Sikorsky S–51s.

The helicopter was already proving its worth in similar circum-stances elsewhere. On the night of Saturday, 31st January 1953, fierce floods swept the east coast of England from Lincolnshire down to Kent, drowning some 300 people, and rendering others homeless or isolated by the rising waters. The following morning, helicopters were in the air to assess the damage and to rescue survivors, providing vital communications and relief in an area in which road and rail transport was severely disrupted and emergency services crippled. There were even more pressing problems from the same floods elsewhere, so that on the second day of the disaster, Royal Navy helicopters were flown across the southern North Sea to the Netherlands, where the floods had inflicted terrible damage on much of that low-lying country, and where they were able to rescue more than 750 survivors.

Two years later, floods struck at Tampico in Mexico, and heli-copters were rushed to the area by the United States Navy, Marine Corps and Air Force, where they rescued what must still be a record 9,262 persons, no less than 2,445 of them by winch, as well as playing the major role in distributing food and medical supplies. The Mayor of Tampico was later to award gold medals to the pilots in appreciation of their strenuous efforts. Also during that year, floods struck close to the helicopter's own birthplace, Connecticut, requiring the rescue by helicopter of more than 1,100 people, and then also on the other side of the United States, in northern California, with about 1,000 people being rescued by helicopters, mainly operated by the United States Coast Guard. On this last operation, one USCG pilot, Lieutenant-Commander George Thometz and his co-pilot Lieutenant Henry Pfeiffer, set a record by rescuing 138 persons, 58 of them by winch, while flying without a rest period for no less than 15 hours!

Possibly one official indication of the recognition being given to the helicopter occurred in Britain during 1953, when the still young machine was seen as fit to appear alongside other more glamorous naval aeroplanes at the Coronation Naval Review on 15th June at Spithead, with helicopters leading a fly-past of 300 aircraft from Royal Navy squadrons to mark the enthronement of the newly crowned Queen Elizabeth II.

One of the first Royal Navy vessels, other than the aircraft-carriers, to be equipped to carry helicopters on a permanent basis was the survey vessel HMS *Vidal*. In October 1954, when Hurricane Hazel struck the Caribbean island of Haiti, *Vidal* and her S–55 helicopter were close at hand and able to render assistance to the unfortunate islanders.

A modern vessel of all-welded construction and fitted with both a helicopter landing platform and a hangar, *Vidal* was later to undertake one of the most unusual missions assigned to any Royal Navy vessel since the end of World War II, indeed, one which smacked rather more of the voyages of discovery of the eighteenth and nineteenth centuries.

On 14th September 1955, HMS *Vidal* sailed under sealed orders from Londonderry in Northern Ireland for the small island of Rockall, a formidable seventy-foot-high rock in the North Atlantic, far off the coast of Scotland, and perhaps best known as a British meteorological area; indeed, that would appear to be its only prior claim to fame. Aboard *Vidal*, apart from her normal crew and the helicopter, were two non-commissioned officers of No. 42 Royal Marine Commando. Four days later, on 18th September, the two marines and *Vidal*'s first lieutenant, with an ornithologist, were landed by helicopter on a tiny ledge at the top of Rockall, where they erected a metal flagstaff and cemented into position a brass plaque commemorating the occasion, which was none other than the formal dedication of Rockall as a part of the United Kingdom. The climb to the top of the rock would have been difficult, if not impossible, had it not been for the helicopter; pounded by heavy seas and lacking any landing place or beach, climbers would have had to start their climb from a small boat, pitching and rolling in a heavy ocean swell and in danger of being broken against a sheer rock face.

The Royal Canadian Navy at this time had also commissioned a new ice-breaker equipped to operate and maintain up to three helicopters, HMCS *Labrador*. *Labrador*'s task was to shepherd merchantmen through the ice of the Canadian Arctic with vital components and equipment for the construction of the DEW-line radar chain, part of North America's air defences and maintained as part of NORAD, the North American Air Defence Command. *Labrador*'s complement of helicopters included a Piasecki HUP and two small Bell HTL–4s, all of which were normally launched at dawn to conduct ice-reconnaissance patrols. However, the helicopters were to prove their worth in roles other than reconnaissance, lifting materials and men to position complete navigational aid stations ashore, including the one at Cape Fisher in the Foxe Channel, on a site selected from the air. The operation started at 08.30 on 8th July, and was completed late the following day, after 28,640 lbs of material had been moved in a total of twenty flying hours by the three helicopters; a remarkable achievement bearing in mind that the HTLs were, of course, none other than small Bell 47s! *Labrador*'s helicopter maintenance team was also later in the same mission to assist in the servicing of helicopters from the USS *Edisto*, an American ice-breaker with provision for carrying helicopters.

The Royal Canadian Navy had formed its first helicopter flight on 30th August 1951, using two Bell HTL–4s, which were soon joined by three plane-guard Sikorsky HO4S–1s in Squadron VH–21, coinciding with the adoption of American-style designations for aircraft and squadrons in the Canadian armed forces. A similar number of Vertol HUP–3 Retrievers were later obtained, and these operated alongside the Bell machines from HMCS *Labrador* and from the light fleet carrier, HMCS *Magnificent*. The first Royal Canadian Navy anti-submarine helicopter squadron, No. HS50, was formed in 1955 at the shore base, HMCS *Shearwater*, using six Sikorsky HO4S–3s, equipped with AN/APQ–7 dipping sonar. Later, this small but assorted collection of helicopters also saw service aboard the Canadian aircraft-carrier, HMCS *Bonaventure*, when she was delivered in 1957, allowing *Magnificent* to return to the Royal Navy, which had loaned her to the Canadians.

Of course, there were other less happy and less innocent, although equally essential, tasks for the helicopter. American reluctance to share nuclear weapons technology during the period immediately after World War II, even though the Soviet Union was rapidly developing its own nuclear weapons and delivery systems, persuaded Britain that an all-British nuclear deterrent was an essential requirement for the long-term survival of the United Kingdom and her allies. Britain then embarked on a programme of, first, atomic weapons development, and then development of a British hydrogen bomb. A programme of tests was arranged, with the first British nuclear weapon being exploded on 3rd October 1952, aboard a disused 1,450-ton Royal Navy frigate, at Montebello Island, off the north-west coast of Australia. This gave the Royal Navy the dual task of keeping would-be intruders and sightseers away from the area before the explosion, and then after the explosion conducting close-quarters reconnaissance of the area. A base ship, HMS *Campania*, an old aircraft-carrier, was stationed in the area, with two Supermarine Sea Otter amphibians and three Westland-Sikorsky WS–51 Dragonfly helicopters, and it was to these helicopters that much of the work fell.

During the next few years, the Royal Navy's helicopters were to find themselves supporting a number of weapons test programmes, although on later tests, the work was often divided between the Royal Navy and the Royal Air Force, with the RAF using English Electric Canberra jet bombers to sample clouds for high altitude nuclear fall-out, while Royal Navy Westland Whirlwind helicopters operated in the survey, reconnaissance and light transport roles. This type of co-operation first appeared at the explosion of the first air-launched British atomic bomb at Maralinga, Australia, where two Royal Navy Whirlwind helicopters were present, and then again on 15th May 1957, for the British hydrogen bomb tests at Christmas Island in the Pacific; code-named 'Operation Grapple'. However, RAF helicopters also eventually appeared on the scene, with two Whirlwinds from No. 22 Squadron ferried to Christmas Island aboard the light fleet carrier, HMS *Warrior*, some little while after the squadron's formation on 1st February 1955, as an air-sea rescue unit operating Whirlwind HAR–2 and HR–14s, initially at Thorney Island on the south coast of England, although later a flight was detached to RAF St Mawgan, near Newquay in Cornwall.

The United States had earlier exploded nuclear devices in the Pacific, at the Marshall Islands, in March 1954, with the Soviet Union following with tests in a remote inland area of Russia during September of that same year. Of possibly far greater significance to the helicopter, however, was the launch on 21st January 1954, of the world's first nuclear-powered submarine, the USS *Nautilus*, 3,500 tons, which was ready for service in November. By January 1956, it was announced that the *Nautilus* had sailed 25,000 miles in a period of 75 days, submerged for half of that distance and time, and in one case spending 3¾ days below the surface, marking the end of the submarine as a small, slow vessel, of limited submerged range ready to surface every twelve hours or so to recharge its batteries. In future, nuclear-powered submarines able to sail more quickly under water than on the surface, and, particularly during bad weather, often able to outpace surface vessels, were to become a major threat to shipping, requiring even greater vigilance and sophistication on the part of anti-submarine forces. The fast and manoeuvrable helicopter, with the detection capability bestowed by the dunking sonar, and able to operate from small vessels near to the scene of potential action, was to become the most important and effective of all the counter-measures.

Other navies had by this time also established a significant helicopter force, possibly as yet another example of military or naval innovation being immediately and effectively countered.

France's fleet air arm, the Aeronavale, operated with leased British and American aircraft-carriers during the years following the end of World War II and up to the late 1950s and early 1960s, until the commissioning of the new light fleet carriers, *Clemenceau* and *Foch*, built in French shipyards and the first purpose-built French-designed aircraft-carriers to enter service, since the *Bearn* of the inter-war years was, of course, a converted battleship. Some of the first French shipboard helicopter operations followed the return from French Indo-China of the US-built light fleet carriers *La Fayette* and *Bois Belleau* during the early 1950s. The early French Navy helicopters included both American and French-built Sikorsky S–55s, and Vertol HUP–2 twin-rotor machines, as well as the ubiquitous Bell 47.

The Netherlands Navy also operated a light fleet aircraft-carrier at this time, the *Karel Doorman*, the former Royal Navy HMS *Venerable*, which in 1948 had replaced the loaned escort carrier hitherto operated by the Dutch Navy, and also called *Karel Doorman*, after the World War II Admiral killed during the Battle of the Java Sea. *Karel Doorman*'s aircraft complement during the 1950s included S–55 helicopters for plane-guard duties.

There were also, of course, the new naval air arms at this time in both Japan and West Germany, formed as old wartime scars began to heal, and as other pressures changed the whole pattern of World War II alliances so completely that once again yesterday's enemies became the allies of today, and vice versa.

The Japanese Maritime Self-Defence Force came officially into existence during 1954, being formed primarily as a coastal anti-submarine force. The first requirement was to train new personnel, including a number of ex-World War II veterans, many of whom became the first generation of post-war instructors after suitable refresher training. Four Bell TH–13 helicopters were amongst the initial equipment, made up almost entirely of training types, including North American SNJ–6 trainers, but the following year the JMSDF also received a number of Sikorsky S–51 helicopters and these were to be joined within a fairly short time by Sikorsky SH–34J; helicopters built under licence in Japan, for anti-submarine warfare.

Lacking the pressures on regional military and naval forces created in the East by the Korean War, it took a little longer for post-war Germany to be allowed to rearm, although in doing so, there was no pretence at a quasi-police or coastguard role, but instead the new German armed forces were assigned to the North Atlantic Treaty Organization. In contrast to the World War II subservience to the Luftwaffe, the new Bundesmarine was equipped with its own Marineflieger, consisting of helicopters and fixed-wing aircraft operating from shore bases in the anti-submarine, search and rescue, maritime-reconnaissance and anti-shipping strike roles. On its formation in 1957, the Marineflieger included a small number of Bristol Sycamore helicopters for training and liaison duties, and these were later joined, as in Japan, by a number of Sikorsky SH–34 helicopters.

Both these new navies, and most of the existing navies, were limited in their ability to operate the helicopter to the best advantages by the absence of warships able to handle helicopters. Only the United States, Britain, Australia, Canada, France and the Netherlands, operated aircraft-carriers during the mid-1950s, although Brazil, Argentina and India were laying plans to do so. The next stage was to consider designing frigates and destroyers able to handle helicopters and maintain these while embarked and away from shore bases for long periods. At the same time, the need remained for a still more powerful helicopter than even the successful Sikorsky S–55, which had brilliantly performed many roles. Sikorsky again provided the answer, the Sikorsky S–58, or, in United States Navy service, the SH–34.

The Sikorsky S–58 was a development of the S–55, designed to meet a United States Navy requirement for a larger helicopter, and one which could offer an all-round improvement in performance, especially while operating from ships at sea. Although retaining the single main rotor and tail rotor layout of the S–55, with the cockpit

The Sikorsky S–58 saw a return to the reverse tricycle undercarriage of the R–5, and went into service with many armies, including the Bundeswehr.

position above the main cabin and the engine installed in the nose, a reverse tricycle undercarriage was fitted, providing far greater stability on a rolling deck and, incidentally, considerable strength. The S–58 was ordered into production on 30th June 1952, and a prototype, designated XHSS–1 by the United States Navy, made a successful first flight on 8th March 1954, powered by a single Wright 1,525-hp R–1820 radial engine. Production orders for the first 258 HSS–1s for the United States Navy were placed before the first flight – another vote of confidence in the manufacturer. The prime task of the HSS–1 was anti-submarine warfare, although the force still had to be divided into dunking-sonar-equipped hunters and torpedo-carrying killers. Eventually re-designated the SH–34G Seabat, another 120 HSS–1Ns, or SH–34Js, were later ordered, with auto-stabilization and additional navigational equipment for night operations. USN deliveries started in August 1955. A single HSS–1F was built and flown on 30th January 1957, using a twin General Electric T58 gas-turbine installation to investigate a feasibility of a gas-turbine version of the S–58 series, although such a machine was never built in volume for military or naval customers by Sikorsky itself.

American-built S–58s were ordered by many foreign navies, including those for the Bundesmarine and the French Aeronavale, and many European and Latin American armed forces, while the German Army, the Bundeswehr also ordered about a hundred troop-carrying versions, and others served with the Thai and Philippino armed forces. Sud Aviation in France built and assembled a large number of S–58s, with about 135 assembled in France from kits of parts, and 165 actually produced under licence.

As before, Westland Aircraft built the S–58 under licence, but introduced a large number of features, including the substitution of gas-turbine propulsion on all British-built machines, whose development and production really belongs to the next chapter.

The S–58 also entered United States Marine Corps service, with the Marines' version, designated HUS–1, ordered on 15th October 1954, with the first deliveries of these twelve passenger troop-carrying helicopters being to Squadron HMR–363 in February 1957. Although intended for the troop-carrying and assault roles, providing battlefield support and transport, this helicopter also played a major part in the development of the helicopter for attack operations, being equipped with rockets and machine-guns in order to provide a punch on its way into enemy-held territory. Four of the USMC HUS–1As were assigned to operate in the Antarctic, in support of the United States research programme, while another five, designated HSS–1, were ordered for the joint US Marines and US Army VIP flight, formed for the use of the President of the United States, and others, in 1960. Altogether, almost four hundred S–58s of different versions were obtained by the United States Marine Corps, with about forty of these being the HUS–1A version, modified for use as an amphibious helicopter by the addition of very bulky inflatable pontoons, and it was essentially this version which was also adopted by the United States Coast Guard as its main search and rescue helicopter.

In United States Army service, the S–58 became known as the H–34A Choctaw, and entered service during spring 1955, but unlike the United States Marines HUS–1, or UH–34D, Seahorses, the Choctaw was built as a sixteen-passenger machine, and was also used frequently as a crane helicopter. Eventually, Sikorsky alone built almost 2,000 S–58s in the various civilian and military applications, and from 1970 onwards offered kits to enable operators to convert their piston-engined machines to gas-turbine propulsion, thus extending their life. Indeed, the S–58 is probably second only to the Bell 47 series amongst Western helicopters for longevity of service life.

Brilliant though the S–58 was, the search for larger and more powerful helicopters was, by the late 1950s, ready to make a considerable leap forward in terms of the number of passengers or the loads which could be carried. A British twin-rotor production helicopter, the Bristol Belvedere was involved in trials aboard the aircraft-carrier, HMS *Eagle*, during the latter part of the decade, in the pursuit of a possible Royal Navy and Royal Canadian Navy order, while the equipment which could be packed into a medium-sized helicopter such as the S–58 was increasing, along with the machine's operational punch. Possibly, even more significant, the helicopter was at last on the verge of seeing ships enter service which were designed to take helicopters to sea in increasing numbers, sometimes using medium-sized helicopters such as the S–58, but more often using small helicopters themselves designed and built with operations from small warships, such as frigates and destroyers, in mind.

5
SUEZ, AND AFTER

By the mid-1950s, as we have already seen, the helicopter had become an indispensable part of the armoury of any air arm of importance, with the major military operators of rotary-wing aircraft able to draw upon the experience of the helicopter in Korea, where it had proven itself as a transport, a flying ambulance or CASEVAC, casualty evacuation, aircraft, and as an ideal machine for aerial observation post, AOP, duties, and on liaison and communications work. In addition, research into the potential of the helicopter had continued in Britain and America, with the military and naval forces exercising with the helicopter almost as a matter of course. The idea of the helicopter as an assault transport, replacing the traditional landing-craft both for the transport of infantry and of weapons and light vehicles, had been conceived and developed by the United States Marine Corps during the early years of the decade, with encouraging results. While serious attention was being paid to the development of larger helicopters on both sides of the Atlantic, and on both sides of the Iron Curtain, air forces and naval and army air arms were able to use their existing helicopters to good effect, including the Sikorsky S–55 and its Westland cousin, the Whirlwind, while production of the improved S–58 got under way.

Unknown to many of those concerned with the helicopter's development, opportunities to demonstrate the machine's suitability for assault operations were beginning to appear. The first of these arose in the Middle East.

Britain had long maintained a presence in Egypt, largely concerned with the need to protect the Suez Canal, which was jointly owned by Britain and France, and through which a great deal of the oil and other raw materials needed by these two countries, and much of the rest of Europe, and the manufactured goods required by countries in the East and in the Pacific region, had to be moved. Indeed, during the late nineteenth and the twentieth centuries, the canal had attained the status of an international waterway, even though it ran wholly through Egyptian territory from the Mediterranean to the Red Sea. The canal was at the peak of its importance to those nations whose trade depended on it, during the first half of the 1950s.

The British military presence in Egypt was steadily withdrawn after a revolution led to the overthrow of the Egyptian monarchy and its replacement by a left-wing nationalist government hostile to the West during the early 1950s. However, Egyptian nationalization of the Suez Canal and the accompanying threat to the free passage of international shipping, created a tense situation in which both the British and French governments found themselves under international and domestic pressure to take action designed to protect the canal's status as an international waterway. The Egyptian presence on the canal in particular meant that Israeli vessels would not be able to use the waterway, and on 29th October 1956, Israel invaded the Sinai Desert, part of Egypt, and fierce fighting ensued between Israeli and Egyptian ground and air forces. Britain and France, which had both mustered strong naval forces in the area over a period of several months, issued an ultimatum to the governments of the two warring nations, requiring both to pull their forces back to positions not less than ten miles from the canal zone, but while Israel complied, the Egyptians refused to do so. These reactions had been anticipated, and the Anglo-French naval force, with a substantial number of vessels equipped for an invasion of Egypt, and supported by carrier-borne aircraft and by British aircraft based in Cyprus, moved towards the canal zone as part of a plan, code-named 'Operation Musketeer', to invade Egypt, launched on 31st October 1956.

At dawn on 1st November 1956, British and French carrier-borne aircraft attacked Egyptian airfields, concentrating initially on those within the canal zone, and these aircraft were soon joined by Royal Air Force bombers and fighters from airfields in Cyprus, although these land-based aircraft could spend little time over the canal due to the constraints placed upon them by the distance from Cyprus. British and French paratroops were flown in, dropping onto Port Said on 5th November, followed later that day by troops from the landing-craft of the invasion fleet. During this initial phase of the invasion, helicopters played an important role, with Sikorsky S–51s and S–55s of the French forces, and British Dragonflies and Whirlwinds, carrying out CASEVAC operations and also being employed on general transport, communications and liaison duties.

An even more significant role for the helicopter was to follow. Early on 6th November, five hundred men of No. 45 Royal Marine Commando were airlifted by helicopter from the two light fleet carriers HMS *Ocean* and *Theseus* to take Gamil Airfield, near Port Said, in the first helicopter-borne assault. Most of the helicopters were Royal Navy Westland Whirlwinds, but these were reinforced by helicopters from a joint Army and Royal Air Force Whirlwind trials unit. Immediately after the successful assault, the helicopters switched to the casualty-evacuation role, with one Royal Marine, injured in fighting after landing with the first wave of troops from the carriers, being returned as a casualty by helicopter and arriving back aboard his ship just twenty minutes after leaving!

The CASEVAC operation by the British helicopters at Suez is often overlooked in the interest surrounding the assault, which was the distinctive feature of this highly controversial operation, which was to have such immense political repercussions in Britain and France. Apart from the normal CASEVAC operation, the Royal Navy Westland Whirlwinds conducted what the United States Air Forces would now describe as combat air rescue, and, in one instance, successfully rescued the pilot of a Royal Navy Sea Hawk jet fighter-bomber who had ejected from his aircraft some thirty miles inland.

The Suez operation was short-lived, being called off within days as, under severe international pressure, the United States threatened to withdraw support for the British pound sterling and the French franc on the international money markets, posing a direct threat to the stability of the British and French economies. Militarily, however, there can be little doubt about the success of the operation, marred only by the time taken, a whole summer, to mount it. Ironically, earlier action by Britain and France might have gained both countries international support while much of the world was still alarmed about the effects of the Egyptian take-over, since the mood earlier in the year had been in favour of strong action, with many countries pressing Britain and France to act.

There was one further novel feature about the Suez operation, although this did not arise until after the withdrawal. On passage to Malta after the withdrawal, the aircraft-carrier HMS *Theseus* was carrying wounded Allied troops, but was running short of vital medical supplies. A Royal Air Force Avro Shackleton maritime-reconnaissance aircraft from the RAF base at Luqa in Malta, flew supplies out to the aircraft-carrier, flying slowly past at just 150 feet above the waves in heavy rain, to drop the containers with the medical supplies, positioning on sea marks dropped by the carrier. As soon as the Shackleton had passed, two Royal Navy Whirlwind helicopters, which were already airborne at the time of the drop, picked up the waterproof containers and had these safely aboard the ship within minutes of the drop being made.

The political, naval and military lessons of the Suez operation were simple. Action had to be immediate if the political and military initiative was not to be lost, and indeed, the term subsequently coined

91

for this type of operation by the British, that of "fire brigade" actions, was most appropriate. This led in turn to the concept of the "commando carrier", transporting a highly mobile amphibious force with landing-craft and helicopters which could be put ashore as advance troops to secure a bridge-head, or to deal with localized outbreaks of violence. Essentially, the helicopter was the new feature, since the Royal Navy had frequently landed detachments of marines from its cruisers from 1900 onwards, and although the British had the example of Suez to guide them, the Americans had already discovered the concept in joint US Navy and Marine Corps exercises during 1952 from the USS *Palau*, described in an earlier chapter.

In 1958, the British Admiralty decided that the Hermes-class light fleet carrier, HMS *Bulwark*, commissioned in 1956, should be withdrawn after her second commission, or tour of duty, ended in spring 1959, so that the ship could be converted to a commando carrier. *Bulwark* was ready for her new role by spring 1960, while one of her sister ships, HMS *Albion*, followed, reappearing as a commando carrier and recommissioning during summer 1962. During conversion, these ships lost their catapults and arrester wires, and gained additional accommodation for a Royal Marine commando unit

The first commando carrier was HMS *Bulwark*, seen here during the early 1960s with some of her Wessex helicopters on the flight-deck.

of six hundred men, which could be increased to 1,200 in an emergency, plus vehicles, landing-craft and, of course, a squadron of transport helicopters. The first Royal Marine unit to be designated for this role was No. 42 Commando, at that time based upon Singapore, ready to embark with guns and vehicles whenever necessary. The helicopter element included sixteen Westland Whirlwinds, with another five in reserve in case of accidents or major mechanical failure, although these were replaced before long by a force of twenty Westland Wessex aboard each ship. In addition to their world-wide commando role, these ships were also well suited to act as disaster relief vessels or to evacuate civilians from a threatened area in an emergency.

The new concept was not to be without a chance to prove itself, showing that in an uncertain world, the concept was just right.

Political instability was nowhere more in evidence than in the Middle East, or at least certain parts of it. Iraq had suffered a revolution, with the overthrow of the Royal Family and the installation of yet another of the region's left-wing revolutionary governments, which in June 1961, claimed sovereignty over the neighbouring state of Kuwait, which, as it happened, had just ratified a new treaty of friendship with Great Britain. Iraqi rebels had earlier ransacked and then burned the British embassy in Baghdad. Against this background of growing uncertainty over its future, Kuwait claimed British protection, and the British Commander-in-Chief Middle East was ordered by the British Government to land troops in Kuwait on 1st July.

HMS *Bulwark* had been on passage to Karachi, in Pakistan, during late June, for a courtesy visit before her final trials as a commando carrier in the Persian Gulf, while one of the Royal Navy's then considerable force of attack carriers, the veteran HMS *Victorious*, which had recently been extensively modernized, was also in the area, *en route* to join the Far East Fleet at Hong Kong: both ships headed towards the Gulf, while a third carrier, one of *Bulwark*'s sisters, HMS *Centaur*, sped to the Gulf from Gibraltar. On schedule, on 1st July, HMS *Bulwark* was in the Gulf off Kuwait, and the men of No. 42 Royal Marine Commando were flown into Kuwait, securing the airfield in temperatures ranging from 90°F. to a maximum of 124°F. The following day, No. 45 Royal Marine Commando was flown into Kuwait from Aden by the Royal Air Force, and joined No. 42 with tactical air support and air cover from the Royal Navy's carriers until relieved by Army units.

The Kuwait incident was a classic example of rapid intervention, and it was completely successful, with hostilities being forestalled by a determined show of strength, preventing not only the loss of life and injury from a war in the Middle East, but also a serious threat to the West's oil supplies.

Even better was to follow. During early 1964, the armies of the newly independent East African states of Kenya, Uganda and Tanganyika, now Tanzania, mutinied, while the government of the Sultan of Zanzibar was also overthrown in another revolution. When units of the Tanganyika Army mutinied, President Nyerere requested British assistance, and the aircraft-carrier, HMS *Centaur*, was dispatched from Aden with No. 45 Royal Marine Commando embarked, although unlike her sisters, she had not been modified to act as a purpose-built commando carrier. *Centaur*'s marines were landed by helicopter directly onto the barracks occupied by the mutineers, finding relatively little resistance and regaining control with three Africans killed and nine wounded. No. 45 was later replaced by No. 41, which had earlier flown direct from Britain to Nairobi, in Kenya, to quell the mutiny there, in response to a similar plea from the Kenyan Government. A detachment of marines also had to be flown to Lake Victoria in Uganda from Nairobi at this time to put down a mutiny by the Uganda Rifles.

In all of these operations, speed was of the essence. The helicopter-borne assault from a suitably equipped warship provided a high chance of success with considerable flexibility of action compared to a paratroop assault, which has to be immediately successful because the troops so landed are in effect cut off for an initial period. By contrast, the command facilities of the warship, the rapid deployment and redeployment of troops made possible by the helicopter, and the close tactical air support from machine-gun or rocket-equipped helicopters, together provided a far more effective answer to such limited but vital campaigns, sometimes also known as "bushfire" wars.

Of course, the chances of success are greatly enhanced by the availability of both the right kind of helicopter and the right kind of supporting warships. Developments were to take place during the early 1960s with both of these requirements in mind.

As we have already seen in the previous chapter, the helicopters were already entering service, with such additions as the Sikorsky S–56 and S–58, and the Vertol H–21, and larger and faster helicopters were under development in the United States. However, the British had also been working on a larger helicopter, the Bristol Belvedere. This was the world's first operational twin-engine tandem-rotor machine and, in its production form, it was a worthy competitor in

94

The need for larger helicopters was satisfied by aircraft such as the RAF's Bristol Belvederes, here ready to lift a Wessex fuselage.

both performance and payload to the American and Soviet machines.

The Belvedere had originated with Bristol's Type 173 design, originally commenced to meet a British Ministry of Supply requirement during 1948. Using two 550-hp Alvis Leonides engines and the rotor heads of the small Sycamore helicopter, the first 173 flew on 3rd January 1952, showing itself capable of lifting its crew of two and up to ten passengers, and this machine took part in trials aboard the new aircraft-carrier, HMS *Eagle*, during 1953. A second prototype was fitted with stub wings, which improved the in flight performance considerably. Although the rotors and engines were interconnected, to provide synchronization and good engine-out capability, the performance of the machine was still not sufficiently powerful for service use, a problem partly due to the non-availability of a suitable gas turbine power plant. After further trials with the up-rated Leonides Major engine, including a version for naval evaluation in 1955, the design settled on the Type 192 Belvedere for the Royal Air Force, of which twenty-four were built, with deliveries starting in August 1961. Two Napier Gazelle turbo-shaft engines of 1,050 hp apiece powered the production Belvedere, which could carry between eighteen and twenty-five fully-equipped troops, or twelve stretcher cases, or up to 5,250 lbs in the helicopter-crane role. Two other

One of the great "might have beens", the Fairey Rotodyne combined the advantages of rotary-wing and fixed-wing flight, with a significant boost in passenger capacity.

A contemporary of the Belvedere was the Westland Westminster, abandoned by the manufacturer in favour of the Rotodyne, before support for the Rotodyne was withdrawn.

versions of the Belvedere, the Type 191 and Type 193, respectively for Royal Navy and Royal Canadian Navy requirements, were never built due to budgetary constraints in both countries. Failure, subsequently, to develop the Belvedere meant that the British helicopter industry failed to capitalize on this worthy machine although, at the time, it was also a helicopter with a limited market, confined to the more important air forces and navies, because of the cost.

Indeed, generally, the British relationship with the helicopter has tended to be less successful and fruitful than with other aircraft types, although there has been success with licence-built Sikorsky S–55, S–58 and S–61 types, and with specialized British-designed machines such as the Wasp and Lynx, and in a share of a number of Anglo-French types. To some extent, the problem has been due to short-sightedness on the part of successive governments, more anxious to support licence-built designs at a lower initial cost, and lower long-term benefit as well. Imagination was certainly not lacking, with a number of designs, from the Cierva Air Horse to the Belvedere, Westminster and Rotodyne, intended to provide a larger helicopter. The Westminster was to be a large helicopter incorporating S–58 rotor heads, but which was cancelled by the manufacturer, Westland, on its acquisition of Fairey and the Rotodyne design in 1960, and this latter aircraft was the lost opportunity *par excellence*.

The Fairey Rotodyne was a compound helicopter of unprecedented size at the time of its first flight on 6th November 1957, having originally been ordered by the then Ministry of Supply, later the Ministry of Aviation, in August 1953. A development of the earlier Fairey Gyrodyne prototypes, which had established a number of British helicopter records, the Rotodyne featured a large rotor powered by air bled from two wingtip mounted Napier Eland turbo-props, using the rotor for vertical take-off and landing, and for hovering, while full power was applied to the tractor propellers of the turbo-props for forward flight. The first flight using the tractor propellers was on 10th April 1958, while on 5th January 1959, the Rotodyne established a helicopter speed record over a closed circuit of 191 mph. The prototype Rotodyne was a three-crew, forty-passenger machine, itself a remarkable achievement for its day, but on the acquisition of Fairey, Westland Aircraft proposed to develop the Rotodyne into a production aircraft capable of carrying between 57 and 75 passengers, and using two of the new 5,250-shp Rolls-Royce Tyne turbo-props to give a cruising speed of 230 mph, and the ability to carry up to 18,000 lbs of freight including standard-width British Army vehicles. However, an initial order for twelve production

Rotodynes for the Royal Air Force did not materialize, and after initial interest from British European Airways, the state-owned domestic service and European international airline, did not develop into a firm order, the project was abandoned in February 1962.

To put the achievement of the Rotodyne into perspective it offered a performance which would have bettered that of the modern conventional helicopter, such as the Sikorsky S–65 series and the Boeing-Vertol Chinook, and while the concept was novel, so too is that of the British Aerospace Harrier jump-jet fighter, which today remains the sole operational Western vertical take-off and landing fighter aircraft. In RAF service, the Rotodyne could have handled the duties of the Hawker Siddeley Andover transport and of the Belvedere helicopter. Helicopter operators and designers should, in any case, be the last members of the aviation community to be frightened by novelty.

Instead, during the late 1950s, the British armed forces, and the Royal Navy in particular, were putting their faith in yet another Westland development of an American Sikorsky design, on this occasion the Sikorsky S–58. However, rather than simply build the American helicopter under licence, Westland improved the performance of the helicopter substantially by substituting a gas turbine engine for the original piston engine. Known in Britain as the Wessex, the licence-built S–58 used an 1,100-shp Napier Gazelle NGa.11 turbo-shaft, itself a licence-built development of the American General Electric GE-T58 turbo-shaft. The new engine was first tried on the Westland-imported Sikorsky S–58 HSS–1 used as a pattern aircraft, and which flew in its modified turbo-shaft-powered form on 17th May 1957. A British-built prototype followed, making its first flight on 20th June 1958.

Production versions of the Wessex for the Royal Navy were delivered from April 1960 onwards, and used a 1,450-shp version of the Gazelle engine, the Mk.161, and were designated the HAS Mk.1 by the Royal Navy, as their first fully-operational 'hunter-killer' anti-submarine helicopter. They were equipped both with dunking sonar and up to two anti-submarine homing torpedoes. After trials with No. 700H flight of the Fleet Air Arm, the Wessex joined No. 815 Squadron in July 1961, and in due course there were six Wessex-equipped anti-submarine and assault squadrons in the Fleet Air Arm. In the assault role, with No. 848 Squadron, for example, the Wessex was able to carry up to sixteen fully-equipped marines or eight stretcher cases. Later versions with up-rated engines were supplied to both the Royal Navy and the Royal Australian Navy, which purchased twenty-seven

Mainstay of British helicopter production during the late 1950s and early 1960s was the Wessex; these four belong to the Royal Australian Navy.

Wessex HAS Mk. 31s. Subsequently, versions of the Wessex built for both the Royal Navy and the Royal Air Force, and for a number of overseas air arms, including the air forces of Iraq and Ghana, used twin-coupled Gnome turboshafts for greater reliability and safety. As well as the later HC Mk. 5 commando-carrying assault helicopter, other versions of the twin-engined Wessex included the machines used by the Royal Navy's search and rescue flights. The assault versions of the Wessex often operated equipped with rocket pods with unguided rockets for ground attack.

Ships proved to be rather longer in coming. The commando carrier concept was excellent, but suffered from difficulties in putting heavy equipment, such as battle tanks and bridging equipment, and heavy self-propelled artillery, ashore. One answer lay in a novel combination of the old and the new, tank landing-ships able to carry helicopters as well as tanks and heavy artillery for the British Army. This was taken a stage further by a class of just two ships for the Royal Navy, the Fearless class, consisting of two 11,060-ton assault ships, fitted with anti-aircraft missiles for close-in defence off a hostile shore, with a helicopter flight-deck aft of the superstructure, but without a hangar, and a special dock at the stern able to carry four LCM(9) landing-craft in addition to four LCVP landing-craft carried in davits. The stern dock flooded with the stern of the vessel dropping to a depth of 32 feet against the usual draught of 20.5 feet, while both ships were designed to carry up to 700 troops, fifteen tanks and thirty other vehicles at any one time. HMS *Fearless* herself was laid down in 1962 and commissioned in 1965, while her sister ship followed two years later. Hardly fast ships, with a service speed of 21 knots, against the 28-knot commando carriers, these ships could handle heavy equipment while also providing all of the space and the communications facilities required for the command role during either a major landing or a more localized 'police' or 'fire brigade' action. Indeed, the ships offered more of a deep sea or 'blue-water' capability than the average tank landing-ship.

In 1956, the year of Suez, the United States Navy and Marine Corps who had already developed the idea of vertical envelopment during Korea, had already decided to convert the USS *Thetis Bay* to be the first helicopter assault carrier, a similar concept to the British idea of a commando carrier, and with accommodation for up to 1,000 combat-equipped marines as well as their helicopters and equipment.

The United States Navy then prepared to equip itself with vessels similar to the Fearless class in concept, although much larger. The USS *Tarawa* and her four sisters displaced 39,300 tons each, and with weaponry boosted by three 5-in. guns in addition to the anti-aircraft

The Americans also experimented with helicopters embarked aboard carriers for assault duties, with this force of Sikorsky S–55s in USMC service.

armament, they presented a massive jump in capability compared to *Fearless* and *Intrepid*. Compared to the five helicopters of the British ships, the *Tarawa*, *Saipon*, *Belleau Wood*, *Nassau* and *Da Nang* can handle up to twenty-six helicopters or even vertical take-off fighter aircraft in lieu of the helicopters, and the complement of troops rises to 1,900 instead of the 700 for the Fearless class. Naturally, the landing dock can handle even larger landing-craft, four LCVs. Laid down from 1971 onwards, the *Tarawa* was commissioned in 1976, and was followed by her sisters, with the final ship commissioning in 1981.

Rather more in the style of the commando carrier, although purpose-built for the role, was the USN's earlier Iwo Jima class of seven amphibious assault ships, each of 18,000 tons and laid down between 1959 and 1969, for commissioning between 1961 and 1970. These single-screw vessels, unusual for aircraft-carrying ships, lacked the full facilities of attack and anti-submarine carriers, but nonetheless could handle up to twenty-four helicopters or vertical and short take-off and landing aircraft, and up to 1,724 troops. These ships were followed in 1970 and 1971 by the USS *Blue Ridge* and *Mount Whitney*, but although having a broader-beam hull, these latest vessels could

The American idea of the purpose-built assault ship, the USS *Iwo Jima*, with an effective variety of Chinooks, Sea Stallions and Iroquois!

In some ways outdated by the H–25, the Piasecki H–21A Work Horse was often used on longer range SAR duties.

only carry one helicopter, being designed for the command role. By comparison, the flight-deck of the Iwo Jima class can handle up to seven large helicopters at any one time.

The late 1950s and early 1960s saw the acceptance of the helicopter reach such an extent that the United States Army, then as now the West's largest operator of helicopters, had no less than half of its 3,600 aircraft accounted for by rotary-wing types. Amongst the newest helicopters at the time were the latest of the Vertol, or Boeing-Vertol line.

The new Vertol H–21 was a distant descendant of the HRP–2 Rescuer. An intermediate helicopter, the ten-seat HRP–2 using a single 500-hp Wright R–340 piston-engine was built solely for the United States Marine Corps, which in fact ordered the staggering total of five aircraft for assault training and exercises. However, the next stage was the Piasecki PD–22, designated the XH–21 by the United States Air Force and Army, for whom a prototype made its first flight on 11th April 1952, before eighteen YH–21 pre-production machines were built for the USAF. The new helicopter showed considerable promise from the start, establishing a number of official helicopter records, including a speed record of 146.735 mph (236.15 kmph) and an altitude record of 22,219 feet (6,795 metres) in September 1953. Deliveries to the United States Air Force commenced in October 1953.

In spite of initial trials as an assault helicopter, the first duties for the United States Air Force's Vertol H–21A Work Horses was the rather more peaceable one of Arctic search and rescue, being capable of carrying up to fourteen passengers, or twelve stretchers in the casualty-evacuation role, in addition to a crew of two. A total of thirty-two machines were built in this form, using a single 1,425-hp Wright R–1820–103 engine de-rated to 1,150 hp. However, the engine was operated at full power in the next version, the H–21B, of which 163 were built for the United States Army as an assault and battlefield transport helicopter capable of carrying up to twenty troops, although later versions of this same helicopter were used by the Military Air Transport Service, MATS, in the search and rescue role, as well as assisting in the support of some of the radar stations in the North American DEW-line chain. By the time the H–21C Shawnee was ready for the United States Army in 1956 as a definitive assault and battlefield transport, Vertol had acquired Piasecki and renamed the machine the Vertol Model 43, although the H–21A and H–21B were known as the Model 42 to the new manufacturer. However, the H–21C was the most widely used of the series, with more than 330 being built, mainly for the United States Army, although a hundred or so were built for the French Army, with another ten for the Aeronavale, sixty or so for the Federal German Army or Bundeswehr, and small numbers for the Royal Canadian Air Force and the Japanese Air Self-Defence Force.

The H–21C was also used in experiments as a flying gunship with machine-guns mounted in the doorway and the undercarriage modified to hold up to four machine-guns or air-to-surface rocket pods, during the late 1950s. The performance of this helicopter also resulted in substantial use in the flying-crane role, but experiments took place, also during the late 1950s, to assess the possibility of using two H–21Cs flying together to lift exceptionally heavy or awkward loads, but there were severe difficulties in control between two helicopters and plans to develop this theme further to use up to eight helicopters at once were abandoned. Other experiments used H–21s modified to take twin General Electric T58 and twin Lycoming T53 turbo-shaft engines, but these were also subsequently abandoned, although in this case the real reason was the advent of the later Boeing-Vertol 107, with turbo-shaft propulsion and an improved performance. A refined and civilianized version of the H–21C was the Vertol 44, originally designated the PH–42, and this was put into production for a number of commercial customers, although a small number were also delivered to the Royal Swedish Navy as the HKP–1 anti-submarine helicopter, with another two for the Japanese Ground Self-Defence Force and three for the Royal Canadian Air Force for the transport role. The Swedish machines were amongst the first helicopters to be built with watertight hulls, in case of an emergency landing at sea.

In United States Army service, no less than ten squadrons were equipped with the H–21C at the machine's peak operational availability, while the same number of squadrons were equipped with the H–34. Nevertheless, the continuous development of the helicopter required manufacturers and operators alike to look ahead, and the mid-1950s saw the United States Army hold a design competition for a new lightweight, turbo-shaft-powered, helicopter, mainly for utility and CASEVAC operations and designated the HU–1A by the Army; the competition was won in 1955 by Bell, and so the ubiquitous Iroquois was conceived.

The United States Air Force meanwhile had been replacing both its Bell HTLs and Sikorsky HO5s with the Kaman HOK, later to be re-designated the H–43A and HH–43B Huskie. The H–43A was used not only as a search and rescue machine, but increasingly as a local crash and fire-fighting machine at USAF bases, while the longer distance or wider radius search and rescue operations gradually passed to the HUS, the USAF's designation for the Sikorsky S–58 in its transport, utility and rescue role.

Although not one of the most significant helicopters, almost 200 Huskies were built in the HH–43B rescue version alone, and a number were exported under the Military Aid Programme, MAP, and the improvement in accommodation offered by the HH–43B, with eight passengers instead of the five of the original version, was increased still later, in 1964, with the eleven-passenger HH–43F, with an improved rotor head and up-rated engines for improved performance in areas with warm climates and airfields at some altitude, with one customer, not surprisingly, being the Imperial Iranian Air Force.

The military and naval use of the helicopter was becoming widespread by the late 1950s, with even such backward and remote countries as Afghanistan managing to operate half a dozen SM–1s, the Polish-built version of the Mil Mi–1, and so too did the politically remote and friendless Albania. Newly independent nations such as Ceylon, now Sri Lanka, operated obsolete Westland-Sikorsky WS–51 Dragonfly helicopters, and Indonesia, already independent for some years, operated Hiller 360s for flying training for the AURI, the Indonesia Air Force.

Of the participants at Suez, obviously the British and French operated helicopters in some quantity, but on the other side, Egypt did not receive significant numbers of helicopters until some time afterwards, during the late 1950s, when a mixture of Soviet and Western types were obtained for the Egyptian Air Force, with a few surviving pre-revolution Sikorsky S–51s being replaced by Polish SM–1s, while one of Egypt's partners in the newly formed United Arab Republic, the Yemen, operated Mil Mi–4s, supposedly Russian-piloted, because the Yemen had great difficulty in finding and training sufficient local pilots, although both Egypt and Syria seemed to be free of this type of problem. The Yemeni helicopters were responsible for directing artillery fire against British forces in the Aden Protectorate in April 1958, one of the earliest examples of this occurring in practice, and indeed, one of the few examples of helicopters being used in this role, reinforcing the belief that Russian air-crew were used for this necessarily skilful task.

Russian helicopter development was also beginning to move ahead at this period, sometimes improving by leaps and bounds and, in certain areas establishing a long lead over even the Americans. However, during the late 1950s, the Soviet Navy had still to develop into a world-wide fleet of great power and strength, a role which was

A flight of Yakovlev Yak–24 'Horse' helicopters in the air.

Yak–24s on exercises, sometimes known to the Russians as the "flying wagons".

not to come until after the Cuban missile crisis of 1962.

The Mil Mi–1 and the larger Mi–4 were destined to remain in production for many years after their introduction, but in 1954, production started of the giant tandem-rotor Yakovlev Yak–24, code-named 'Horse' by NATO, and, of course, designed by Alexander Yakovlev. At the time, this was the largest helicopter in the world, capable of lifting up to forty troops, or two armoured vehicles or three staff cars, or two anti-tank guns, with their crews. Designed specifically for the assault role, the Yak–24 replaced Yakovlev-designed assault gliders in Soviet Frontal Aviation service and provided the capability for the Soviet armed forces to mount an airborne assault superior to that of the British efforts at Suez. The Yak–24 soon lost its title as the largest helicopter to the Mil Mi–6, or 'Hook' in NATO terminology, which made its first flight during late 1957 and was the first practical Soviet turbine-powered helicopter. However, it is not thought that production of this very successful machine started before 1960, although prototypes established a number of records, including speed and height records with payloads, and indeed these were to be improved upon again in 1962 and in 1972, when a Mi–6 took a load of ten tons to an altitude of 16,000 feet. The Mi–6 had the rotary-wing lift augmented by stub wings, although this feature was omitted in the flying-crane version. Passenger accommodation in the Mi–6 is reputed to reach 120 persons, although with fully equipped troops, the figure falls to a more practical 65 persons, while in the CASEVAC role, up to 41 stretchers can be accommodated. Apart from the obvious assault role, during combat conditions such a large helicopter must have an important part to play in bridging, or in the recovering of damaged aircraft and vehicles, including other lesser helicopters, since the 1962 heavy lift record for this machine was no less than 20 tons, heavier than other medium-sized helicopters such as the Sikorsky S–61 and S–64!

Although the Yak–24 remained the preserve of the Soviet Union's own armed forces, and did not appear to have been built in very large numbers, the Mi–6 was supplied to many Russian 'client' air forces, such as Bulgaria, Egypt, Indonesia before the fall of the dictator, President Soekarno, and even North Vietnam, although surprisingly, the last-mentioned did not make very much use of the helicopter during the Vietnam War.

Of course, not all military aid was provided by the Soviet Union, and after Laos was invaded by Viet Minh troops during spring 1953, a US Military Advisory Group visited neighbouring Thailand to assess the needs of this peaceful but strategically important south-east Asian country, while afterwards four Westland-Sikorsky WS–51s were supplied from Britain, followed by a gift of Sikorsky S–55s and the small Hiller 360 training helicopters from the United States.

The Spanish-American Defence Treaty of 26th September 1953, was also designed to ensure that the United States supplied modern weapons to Spain in return for the use of Spanish port and airfield facilities by the United States Navy and Air Force, especially while engaged on NATO duties. Naturally, helicopters figured prominently in this military aid, and the Spanish forces received Sikorsky H–19 Chickasaw and Bell 47G helicopters, with the latter being used exclusively by the Navy.

Neutral countries also obtained helicopters. Late in the 1950s, Switzerland acquired Aerospatiale Alouette II and Sud Aviation Djinn helicopters from France, while Austria, after the withdrawal of occupying forces, restarted her armed forces with three Bell 47G helicopters in 1956 to equip an air force training flight, while six Westland Whirlwinds followed in 1957 to form the first operational squadron, which was soon joined by a second squadron, but this time equipped with Sud Aviation SE.3130 Alouette IIs. While Sweden's first helicopters were the four Vertol 44s already mentioned, the Air Force and Army soon followed the Royal Swedish Navy into helicopter operation, although the RSwN took the responsibility for overall helicopter procurement and pilot training for all three Swedish armed services.

France was the first country outside the Soviet Union and the United States to design and build helicopters in significant quantity and with consistency, compared to the occasional British success. French involvement with the helicopter after World War II started with the Sud-Est SE.3101 single-seat helicopter in 1948, while Sud Aviation also built a small helicopter, the SA.3110 before proceeding to a development of this machine, the experimental 3120 Alouette, or Lark, and then to the 3130, the Alouette II, which first flew on 12th March 1955 using a single 530-shp Turbomeca Artouste II turbo-shaft, and which established an altitude record of 26,937 feet, or 8,200 metres, in June 1955. Eventually, no less than 363 Alouette IIs were built for the French armed forces alone, with most being delivered to the Army, while the new German armed forces received 267 Alouette IIs, with most of these 115-mph, five-seat machines also passing into Bundeswehr service. Other operators included the British, Belgian, Swiss and Swedish armed forces, and a large number of neutral or non-aligned countries. No doubt, the availability of a small mass-produced helicopter of relatively high performance from a supplier other than the Soviet Union or the United States was part of the appeal of the Alouette to these customers!

The Alouette II was not the only French turbine-powered helicopter, with Sud-Ouest building the SO.1221 Djinn, which unusually worked on the so-called "cold jet" principle, using turbine efflux ducted through to exhaust at the rotor-blade tips. First flown on 2nd January 1953, the Djinn was a small two seat machine, with an initial twenty-two being built, nineteen of them for the French Army and three for the United States Army as the 6YHO–1DJ. Eventually, more than a hundred Djinns were built for the Aviation Légère de l'Armée de Terre, or French Army Aviation, and the machine figured prominently in early trials of the Nord AS.10 air-to-ground anti-tank missile.

The Alouette II eventually led to a larger development, the Alouette III, and to a hybrid with features of both the Alouette II and III, known as the Lama, playing its part in maintaining a prominent position for French helicopters in international markets to the present day.

During the late 1950s, France was also one of the countries using its helicopters on active service. War had broken out in Algeria, with French nationals who had settled in Algeria opposing independence, and Algerian Nationalists who sought independence for their country, fighting each other and both fighting the French armed forces, who were in the unenviable position of standing between the opposing factions in a full-scale civil war with considerable consequences for the French domestic political situation, and for the French armed forces, which suffered a mutiny at one point. No less than three Escadres d' Hélicoptères, or helicopter wings, operated in Algeria as counter-insurgency forces, with an inventory of helicopters which included Sikorsky S–55s and S–58s, Bell 47Gs and the then new Sud Aviation SE.3130 Alouette II. Indeed, during 1958, out of 850 French military aircraft operating in Algeria, there were 19 Bell 47Gs, 15 Sikorsky S–55s, 44 Sikorsky S–58s and 19 Alouette IIs.

Post-Suez, the Aeronavale operated three helicopter squadrons, including Squadron NG 31K with Vertol H–21s, 32F with HRS–3s and 33F with S–55s.

The French Army's Aviation Légère de l'Armée de Terre, or ALAT, operated some 600 fixed-wing aircraft and 270 helicopters by 1958, including the Djinn and Alouette IIs, along with Bell 47Gs, Sikorsky H–19s and Vertol H–21s, while a thriving Centre d'Instruction de Specialistes de l'ALAT had been established at Essey-des-Nancy for mechanics and pilots to receive operational training.

Europe's newest armed forces, those of West Germany, had also taken to the helicopter from the start. Established as a result of the Treaty of Paris on 10th April 1955, a headquarters organization was in existence by the following February. The new German Army or Bundeswehr received fourteen Bell 47Gs in 1957 to train helicopter pilots, who later went on to fly one of the 26 Sikorsky H–34A Choctaws or 26 Vertol H–21Hs Shawnees available for assault duties, or the 50 Bristol Sycamore 171 Mk. 14 for CASEVAC, search and rescue and communications duties, or the general-purpose fleet of 14 Hiller UH-12Cs, 11 Saro Skeeters or the Sud-Ouest SO.1221 Djinns. However, at the outset, the new Federal German Navy still used fixed-wing amphibians, the Grumman SA–16A Albatross for search and rescue.

The helicopter was indeed spreading across the globe, with all but the poorest nations operating helicopters by this time, and many of these did not have to wait very long, either receiving military aid or obtaining obsolete types on the second-hand market. To be fair, no one could really afford to be without at least a small helicopter element, so useful and invaluable had the helicopter become.

6
COUNTER-INSURGENCY

In a sense, counter-insurgency or counter-revolutionary activity is as old as the concept of nationhood itself and, in addition to serious counter-insurgency operations, there has at different places and at different times also been the need to counter bandit activity or engage in anti-smuggling duties, as well as patrol areas of frontier or coastline to discourage illegal immigration. Nevertheless, since the end of World War II, such activities have acquired a new importance, largely because of the concerted efforts of extreme left-wing political elements which have introduced sinister supra-national overtones to what has in the past appeared to be the accepted, and even one sometimes suspects, acceptable, means of changing government in some of the less democratic parts of the world. It can even be argued that it is not the number of revolutions or attempted revolutions which is such a serious feature of the post-war world, but rather that so many give the impression of having been financed and organized by an outside force, while the influx of modern weaponry on both sides has helped to ensure that the conflicts are longer lasting and bloodier than most of their pre-World War II counterparts.

Even if it were nothing else, the helicopter has shown itself to be the supreme counter-insurgency weapon. Troops and police can be moved quickly and effectively, while attack from the relatively slow-moving helicopter is so much more effective against guerillas than anything by a jet fighter or fighter-bomber could be, and certainly less costly. Very heavy and sophisticated weaponry is wasted on counter-insurgency, or COIN, operations, high-performance jet aircraft are too expensive, and even the smallest fixed-wing light aircraft needs a runway of some kind, which is no easy task in jungle or bushland. There is a more positive side to this too, for every truly effective counter-insurgency operation should also be concerned to retain the "hearts and minds" of the indigenous population, making friends so that they deny hospitality and refuge to terrorist groups: part of every "hearts and minds" operation includes providing improved medical aid, with doctors or nurses being flown in, or very ill patients being flown out to a central hospital. For the anti-terrorist forces, CASEVAC is also important in maintaining morale as well as for the obvious humanitarian reasons.

111

While the helicopter has been so closely associated with many post-war COIN operations, it should also be remembered that it was absent from several of the early post-World War II operations. There were no helicopters in the Dutch East Indies or in French Indo-China as these colonial powers struggled, unsuccessfully, to reassert their authority, or at least to hand over power in an ordered manner. The helicopter was also absent during the emergency in Palestine, which led ultimately to the formation of the state of Israel, and nor was it present in any significant numbers during the emergency in Kenya when British troops and airmen struggled to combat Mau Mau terrorism. During the campaign against Communist bandits in Malaya, the helicopter appeared only in limited numbers, with at first just a flight of Royal Air Force Westland-Sikorsky WS–51 Dragonflies, which was later supplemented by a squadron of Royal Navy Westland Whirlwinds. Nor did the helicopter figure prominently in the civil war in Cuba, which led to the take-over of the country by Fidel Castro, even though this occurred during the late 1950s!

One of the first major counter-insurgency operations in which the helicopter was to play an important role was in Cyprus during the late 1950s. This was a colonial war in which, as sometimes happened, independence was not the main issue, with the complication and the bloodshed arising from the desire of certain Greek Cypriots for "Enosis" or union with Greece, contrary to the independence planned by Britain for the island, and accepted, with some misgivings, by the Turkish Cypriot minority. This led to a long campaign of violence, with British troops and Turkish Cypriot police co-operating to retain control of the situation in this island of some 600,000 people, with an attractive and varied landscape, including two major mountain ranges.

Throughout most of the emergency, helicopter support for the security forces was provided by Bristol Sycamores of No. 284 Squadron, Royal Air Force, cited by one governor of Cyprus, Field Marshal Lord Harding, on his departure in 1958, as having contributed "more to fighting terrorism on the island than any other single unit". By this time, the squadron had spent two years on the island, and during 16,000 sorties had dropped 3,500 troops on terrorist-frequented countryside, trained 13,000 troops in the then novel skill of scrambling down ropes from helicopters, evacuated 200 casualties, and dropped 113 tons of ammunition, food and other supplies to troops on counter-insurgency duties. There was a considerable amount of pioneering work in this, including night flying, which was still a little-used art in helicopter operation at the time, and, at first unique to this squadron, the dropping of troops in mountainous terrain and in forest, creating the need for troops to learn to scramble down ropes, reducing the risk of rotor blades striking trees or rocky outcrops.

One of 284 Squadron's more notable operations included that at the Makheras Monastery, a terrorist hideout 3,000 feet up in the Troodos Mountains, some twenty miles south-west of Nicosia. It would have been difficult to mount this operation successfully without helicopters, five of which flew in forty troops to seal off the mountain roads while other troops moved in on the ground. Gregorious Afxenthiou, Chief of Staff to the EOKA terrorist organization's leader, Colonel George Grivas, was killed in the operation while other terrorists were captured.

There was another side to 284's operations in Cyprus, including assistance in fighting forest fires and search and rescue operations off the coast of the island. However, it was the Sycamores of another RAF squadron, No. 103, which were involved in one of the most important air-sea rescue operations during the period. During a heavy storm in 1960, a Yugoslav freighter, the MV *Snjeznik*, got into difficulties while attempting to tow another vessel in trouble, the Japanese MV *Nagoto Maru*, which had run aground. Three of 103 Squadron's diminutive Sycamores flew to the *Snjeznik*, and one helicopter dropped a French-speaking RAF officer to brief the crew on the rescue procedure before all thirteen crew members were taken off. Unfortunately, in the gusty conditions prevailing at the time, one helicopter approached too close to the rigging of the vessel's masts, damaging a rotor blade and falling into the sea, although fortunately this potentially disastrous accident had a happy ending, with the escape of the helicopter's pilot, who was winched to safety by another of his squadron's helicopters.

At a rather more mundane level, Royal Navy helicopters in Malta at this time were used to monitor demonstrations and strikes which swept the Maltese islands during the late 1950s before independence. However, there was no outbreak of terrorist activity in Malta.

The next major operation to involve British forces took place in the Far East, and could not be accurately described as an counter-insurgency operation. The main event, a confrontation between the newly independent state of Malaysia, which at that time included Singapore, and Indonesia, which opposed the Federation of Malaysia and also entertained territorial ambitions in Borneo, was initiated by Indonesia, which was also behind a number of localized outbreaks of violence.

On 8th December 1962, a large-scale revolt broke out in Brunei and in parts of neighbouring Sarawak and North Borneo adjacent to

Brunei, as the so-called North Brunei Liberation Army fought against Brunei's inclusion in the Federation of Malaysia. The Sultan of Brunei appealed to Britain for help in maintaining order, following which advance units of No. 42 Royal Marine Commando were flown in from their base at Singapore, and after landing these troops moved inland using requisitioned lighters as transport to rescue European hostages, while the rest of the commando which followed took up the chase after guerillas using river boats and helicopters.

Many of the helicopters used in the Brunei campaign were those of No. 66 Squadron, Royal Air Force, whose Bristol Belvedere helicopters made one of the longest non-stop over-water crossings by a large number of helicopters at that time; flying 400 miles across the South China Sea from Singapore, their main base, to Kuching in Sarawak, before refuelling for the 500-mile flight over dense forest and mangrove swamp to Labaun Island off the coast of North Borneo. Of course, this was an empty positioning flight, and most of the troops taken to Borneo travelled by sea aboard the recently converted commando carrier, HMS *Albion*, which had been exercising in the area, but which was diverted at top speed to Singapore to pick up the headquarters staff of No. 3 Commando Brigade and the men of No. 40 Commando, Royal Marines. The men of No. 40 Commando were landed by HMS *Albion*'s Westland Whirlwind helicopters to join No. 42 Commando, after which the RAF Belvederes and the Royal Navy Whirlwinds operated in support of the marines, airlifting troops and supplies, transporting prisoners and undertaking CASEVAC duties, and then providing assistance and relief for the local population after severe weather struck the area. *Albion* was later joined by her sister ship, HMS *Bulwark*.

It took just one month to disperse the insurgents, proving once again the value of rapid and effective intervention.

After the Federation of Malaysia formally came into being in September 1963, Brunei was not included after a dispute over possession of the state's oil resources. The new federation was immediately faced with a violent reaction from Indonesia, whose dictator, President Soekarno, threatened confrontation, including a continuation of the effort started by the North Brunei Liberation Army, which he had supported. By February 1964, RAF and Royal Navy helicopters, including some of the new Westland Wessex, were operating from bases in Sarawak and Sabah to assist Army and Marine detachments fighting guerilla forces infiltrated by Indonesia over its one thousand mile frontier with Malaysia. Anti-terrorist forces in the area also included units from the Brigade of Ghurkas while Australian

114

and New Zealand forces were also deployed to assist British and Malaysian forces. During the confrontation, Royal Navy Wessex helicopters set records for operational flying over the steaming jungles of Borneo, carrying troops and supplies in all weathers, and doing much to refine COIN techniques during the period up to the end of confrontation, in mid-1966 after Soekarno's overthrow by moderate elements amongst the Indonesian leadership.

An important part of the campaign was the "hearts and minds" element, to win the support of the local population, providing supplies and medical assistance, and in this the helicopter also proved its worth. At no time did the Indonesia terrorists find widespread support amongst the Malay villagers.

After a successful campaign, British forces started to withdraw in September 1966.

Events then centred on Aden. In the pre-independence period, opposition to the South Arabian Federation, a confederation of small states with British support, was backed by the neighbouring Yemen Arab Republic, where a revolutionary government opposed the British plans. Yemeni attentions soon turned to a take-over of the federation, and in turn Yemeni ambitions were supported by Egypt, at that time under a pro-Soviet, anti-Western leadership. The Federal Army was reinforced by British Army and Royal Marine units, with air support provided by Royal Air Force and Royal Navy helicopters. Although there were border skirmishes with Yemeni forces, the main effort was undertaken by two terrorist organizations, the Egyptian-backed National Front for the Liberation of Occupied South Yemen, or NLF, and the Front for the Liberation of Occupied South Yemen, or FLOSY. These groups supported, and may even have instigated, a rebellion by tribesmen in the Radfan Mountains, to the north of Aden, in 1964. In spite of a tremendous effort by the British forces, supported by Royal Navy helicopters from HMS *Albion* and the RAF base at Khormaksar, where RAF helicopters were also based to support the ground forces, the situation remained difficult up to the granting of independence and withdrawal from the South Arabian Federation at the end of November 1967, following which a revolutionary government swept to power. The failure of the British effort in Aden and the neighbouring federation territories can be probably traced to a lack of the necessary will by the British Government, which had determined to leave the territory, rather than to any tactical mistake by the security forces. In any COIN operation, adherence to a fixed date for withdrawal strengthens the position of the insurgents, while doing nothing for the morale or the motivation of the security forces, particularly those drawn from the local population who must obviously be concerned about their future, and that of their families, after the withdrawal of outside forces.

The steady reduction in the size of the British forces stationed outside Europe during the late 1960s and early 1970s accelerated with time, in spite of misgivings amongst many in British political life and of many outside Britain, not least amongst the governments of the smaller and more vulnerable colonies. The loss of the British presence was strongly felt in the Caribbean, in South-East Asia and in the Persian Gulf. In all three areas, Communist-backed forces from a local Russian client state threatened the security of the area. Post-revolutionary Cuba threatening the stability of many Caribbean states, while Iraq threatened the security of the Gulf states, although these relatively wealthy countries were able to equip themselves with sufficient modern equipment to deter any aggressor. In South-East Asia, the growing Communist influence which had followed the establishment of a Marxist regime in North Vietnam, threatened the stability of the entire region. However, the prevailing mood in Britain was against a substantial permanent military and naval presence "East of Suez", and a date was set even for the British withdrawal from Singapore in 1971, after which the sole outside presence in this area would be provided by Australian and New Zealand troops, with support if necessary from British forces stationed in Hong Kong, the one exception to the plans for withdrawal. One of the last British units to leave Singapore was an RAF helicopter squadron with ten Westland Whirlwind helicopters for SAR and transport duties, and indeed, the withdrawal of this unit did not take place until its disbandment in 1974, by which time the Whirlwinds had been replaced by the more powerful and modern Wessex.

Before the withdrawal from Singapore and Malaysia, a major exercise was held to demonstrate the ability of British forces to reinforce the region's own forces in an emergency. Named 'Bersatu Padu', the exercise included the airlift of RAF Westland Wessex helicopters of No. 72 Squadron, from RAF Odiham to Malaya, where they in turn transported troops of the 9th Royal Malaya Regiment to the "front line", which was in fact spread across jungle-covered hills and low mountains. Once the Malayan troops were in position, the RAF helicopters then switched to the support role, keeping them supplied with food and ammunition throughout June 1970. Altogether, a total of twenty RAF helicopters participated in the exercise, although much of the point of the operation was lost within a few years when the withdrawal, for reasons of economy, of the RAF's

116

Short Belfast heavy freighters left the service without the capability to transport other than small helicopters using its own aircraft. This problem was to hamper a real operation some years afterwards, of which more later.

Another element in the post-withdrawal support for the region was in the secondment of British officers and NCOs to the Royal Brunei Malay Defence Regiment in North Borneo, in effect a unified defence force maintained by the Sultan of Brunei rather than simply, as the name implies, an army unit. This arrangement continues to the present, with seconded personnel reinforced by contract personnel, former British servicemen on contract to the Sultan. In 1971, one group of RAF personnel on detachment operated a flight of Wessex helicopters, one of which was sent, in less than ideal flying conditions, to assist the crew of an oil-drilling barge on fire off the coast of Sarawak. The pilot of the Wessex, Master Pilot Alexander Riddoch, flew 180 miles in darkness and torrential rain to the barge, ablaze and rolling in an eight-foot swell, with the task of rescuing the crew made more difficult and hazardous by the collapse of the oil-drilling derrick and numerous trailing wires and guidelines. In these difficult circumstances, Riddoch was entirely dependent on instructions from his winchman, Sergeant Richard Birley. A third member of the Wessex's crew was winched to the deck of the barge, and despite the fumes, heat, smoke and the motion of the barge, several injured crew members were taken off and given first aid aboard the helicopter. Birley was later to fly on four further rescue missions to the barge during that day, actually lifting eight crew members himself. For their efforts, Alexander Riddoch was awarded the Air Force Cross and Richard Birley the Air Force Medal, the highest awards for British airmen in peacetime.

Amongst the other air arms benefiting from the secondment of British personnel was the Sultan of Oman's Air Force, not far from the former troublespot at Aden and actually bordering upon the vital entrance to the Persian Gulf. Here, one-third of the Sultan's pilots were RAF personnel on secondment, some of then flying Bell 205 Iroquois helicopters which could carry up to thirteen armed troops at a time on patrols along sections of the small state's 1,000-mile-long frontier and coastline, on the alert for Yemeni infiltrators.

Not all anti-terrorist operations could be regarded as successful, and one of the less satisfactory roles was placed upon the Royal Air Force in September 1970. Leila Khaled, an Arab terrorist who had failed in an attempt to hijack an Israeli airliner, had been imprisoned by the British when the aircraft landed at London, but a VC.10 airliner of the

British Overseas Airways Corporation (now British Airways), had been hijacked and its passengers held hostage in return for Khaled's freedom. Eventually, an exchange was agreed, and an RAF helicopter and its crew suffered the doubtful privilege of flying the terrorist from RAF Northolt, just outside London, to RAF Lyneham, where she was put aboard an RAF de Havilland Comet VI and flown to the Middle East.

On the other hand, not all COIN activities consist of maintaining the peace. Often a policing role means just that, vigilance and deterrence rather than violence. The one exception to the run-down of the British presence east of Suez was in Hong Kong, a British colony in which a flotilla of naval vessels, several British Army units, usually including a Ghurka battalion, and an RAF Wessex helicopter squadron, are based, officially to assist the Royal Hong Kong Police in maintaining law and order in the colony. Although there were some Chinese Communist-inspired riots during the late 1960s, by and large the colony is peaceful. However, the British forces in the overcrowded, but prosperous colony, are usually busy searching for illegal immigrants from Communist China. The helicopters usually support troops operating along the border between the New Territories leased from China and the People's Republic itself, their availability ensuring the rapid movement and redeployment of troops and police, while the small size of the colony puts the helicopter at a complete advantage compared to any other type of aircraft. The role includes ferrying in supplies to troops in the field, CASEVAC, although usually due to accidents rather than fighting, and the transport of captured illegal immigrants, who are questioned before being handed back to the Chinese authorities.

Concurrent with many of these events, war had been raging in Vietnam. Infiltration of guerilla forces by Communist North Vietnam into the South, and into neutral Laos and Cambodia, had threatened the stability of South-East Asia, and the military assistance and military advisers provided by the United States during the early 1960s soon developed into a whole-hearted major military and naval commitment, with American Army and Marine units fighting the Vietcong troops alongside South Vietnamese military formations, while the United States Air Force and United States Navy and Marine Corps aircraft attacked Vietcong positions in South Vietnam, Laos and Cambodia, and strategic targets in the North, with patrols provided offshore by the United States Navy and Coast Guard Service, including battleship bombardment of enemy shore positions. By the middle of the decade, the growing American involvement was joined

118

Typical of the scene above Vietnam, with Bell UH–1 Iroquois helicopters on patrol, and an infantryman waiting to go into action.

American troops wait while their Bell Iroquois takes off.

by combat units from Australia, New Zealand and South Korea.

Vietnam marked the operational debut of another classic helicopter design, indeed, one of the most popular helicopters built for military users, and certainly a machine which could be regarded as a genuine workhorse, able to undertake a wide variety of missions and operate at full stretch reliably and for long periods, the Bell 204 and 205 Iroquois.

The Bell 204 had its origins in a design competition held some years before Vietnam became an issue, to find a new utility helicopter for the United States Army, with the ability to cover several roles, including light transport, liaison and casualty evacuation, as well as training. Three prototypes were ordered in June 1955, with the US Army designation of XH–40, and the first flight of one of these machines, powered by a 700-shp Lycoming T53 turbo-shaft, occurred on 22nd October 1956. The 204 broke away from earlier helicopter design in that it could accommodate two crew and up to seven passengers, not far short of the load of the S–55, in a configuration more akin to that of a small helicopter, with the cabin immediately behind the flight deck or cockpit, and below the engine. Large cabin doors enabled passengers, often combat troops, to enter and leave easily, although headroom was not over-generous in this squat, compact, helicopter. Successful trials led to an order for six development YH–40s, using slightly more powerful 770-shp variants of the original Lycoming engine and incorporating a number of modifications, before the US Army ordered nine pre-production machines, designated HU–1, although this was later changed, in 1962, to UH–1, and the first of which was delivered during June 1959. It was the HU designation, although soon dropped, which led to the popular nickname of 'Huey' for all of the Iroquois series, indeed, this has been used more frequently than the official Iroquois name, or the US Army or the manufacturer's designations.

The first of many production machines were the seventy-four HU–1A Iroquois for the United States Army, and which were followed by the higher performance 960-shp Lycoming T53–L–5 powered YHU–1B, the pre-production version of some 700 HU–1Bs. Although used for US Army trials of the French Nord AS.11 wire-guided anti-tank missile, the HU–1B was more usually equipped with machine-guns or unguided rockets, or even a combination of both, and it was in this form that the first HU–1Bs arrived in South Vietnam during autumn 1962. The usual armamemt for the Iroquois operating in Vietnam included up to four 7.62-mm M–62 machine-guns and two pods of air-to-ground unguided rockets.

Difficult bush conditions prevent this Iroquois from landing on its errand of mercy; one of the troops on the ground is helping to steady the helicopter.

Panic, as a wounded Korean infantryman clings desperately to the skid of an Iroquois as it takes off.

Further performance improvements came with the UH–1C, using an up-rated 1,100-shp Lycoming T53–L–11 turbo-shaft, and with a new Bell rotor of rather more conventional, hinged, design than that used previously on the company's helicopters. However, a further step forward came with the Bell 205, designated the UH–1D by the United States Army, which entered service in May 1963, and while using the same power-plant as the UH–1C offered far greater accommodation in a stretched cabin, capable of carrying up to twelve fully-equipped troops or six stretcher cases; and more than one thousand examples of this variant were built. Rather confusingly, the next version, the UH–1E, retained the cabin of the 204 series, also entering both USMC and USAF service as the utility UH–1F and training TH–1F, while a later helicopter of similar performance and cabin configuration was the UH–1L, built for the United States Navy along with a number of TH–1Ls.

A more obvious successor to the UH–1D was the UH–1H, which retained the longer cabin of the 205 series and combined this with a still further up-rated Lycoming T53–L–13 turbo-shaft of 1,400 shp, exactly double the power output of the engine used in the early prototypes! Built in large numbers for the United States and allied armed forces, the UH–1H was also built under licence in Japan by Fuji. Yet further development, amongst many, of the 205 series, was the UH–1N, using a 1,250-shp PT6T Turbo-Twin Pac engine for greater reliability, and built by Bell for both the American and Canadian armed forces.

The popularity of the 204 and 205 led to their being built in large numbers, outside the United States, notably by Bell's main European licensee, Agusta, with the Italian company building both the 204 and the 205 as the AB.204 and AB.205. First flight of the AB.204 was on 10th May 1961, using a 1,050-shp Bristol Siddeley (later Rolls-Royce) Gnome turbo-shaft engine, although later versions also used a 1,250-shp version of the Gnome, and both Lycoming and General Electric engines have also been available on Agusta-built 204s and 205s. Most Bell-built export versions were designated 204 or 205 rather than using the American military designations, and these were the same designations applied to the growing number of civil versions. While the United States Navy did not use its Bell Iroquois for anti-submarine duties, Italian-built versions included a dunking-sonar anti-submarine variant for both the Italian and Spanish Navies, designated the Agusta-Bell AB.204AS.

Vietnam was the helicopter war *par excellence*. While bombers, fighter-bombers and transport aircraft all had an important role to play in the war, the helicopter was the star, being by far the most numerous aircraft type and playing the most important roles. Helicopters were used for machine-gun and rocket attacks on enemy forces, as well as for transport, supply, assault, CASEVAC and forward air control, helping to direct American gunfire onto Vietcong positions, although for forward air control of fighter-bombers, the USAF for long clung to a force of Cessna light aircraft, on the basis that these were quieter, lighter and able to loiter longer in forward positions: there was also a shortage of helicopter pilots at different times, and this forced some serious consideration of the role of the helicopter, so that they were allocated the most important tasks.

Every day, helicopters would be ferrying troops between battle positions as the American and Vietnamese forces struggled to maintain contact with fast-moving guerilla units skilled in jungle warfare. Before the war was far advanced, the type of tactics seen in Korea and at Suez became an everyday affair, even a several times a day occurrence, and indeed, the number of troops moved daily by helicopter were far in excess of anything in these earlier operations.

Amidst the daily routine, however, a number of specific operations stood out, and so too did the performance of the Australian units in South Vietnam.

The Australian presence in South Vietnam was the most significant after that of the United States. No less than one-fifth of Australia's small army fought in Vietnam, some 8,000 men at the peak and supported by both Royal Australian Air Force and Royal Australian Navy Fleet Air Arm units, with the latter's commitment consisting entirely of helicopters in South Vietnam, while the RAAF contribution also included a significant proportion of helicopter units. Australian troops impressed observers with their professionalism, preferring to fight the enemy on his own ground inside the jungle rather than outside it. One commentator made the point that the Australians did not bulldoze or blast the rubber trees, preferring instead to camp amongst them for shade and cover, while even the officers did most of their own laundry, rather than employ Vietnamese civilians who might be Vietcong agents, and while on patrol, there were no transistor radios or tape recorders to announce their presence. Not surprisingly, senior United States officers rated the 'Aussies' highly, preferring to deploy them in areas where their abilities, borne out of experience fighting alongside British units in Malaysia, could be used to the full, although this was not possible because of the strong opposition to the Australian presence in Vietnam of many Australian politicians. However, the Australians undoubtedly had a high degree of control over their own areas, working not on a kill ratio, but on keeping the enemy moving, and wearing him down.

Australian forces first arrived in South Vietnam in 1966, their numbers reaching a peak shortly afterwards in 1967–8, with a RAAF Iroquois squadron being amongst the early arrivals, although from time to time United States Navy helicopters were also used until the Australian presence built up. The elderly aircraft-carrier, HMAS *Sydney*, was pressed into service as a fast troop transport, carrying equipment and helicopters to Vietnam, while the light fleet carrier HMAS *Melbourne* operated in Vietnamese waters, although as already mentioned, no RAN fixed-wing units operated out of Vietnamese airfields, and *Melbourne*'s Douglas A–4 Skyhawks were not in the event used extensively, and all of their operations were in support only of Australian ground forces. Bowing to political pressures inside Australia, the Australian presence in South Vietnam was reduced after January 1970, at first by 2,000 troops, but the remaining 6,000 men were soon recalled, although the helicopter squadrons were the last to leave, with the United States, and even more so the South Vietnamese, raising strong objections to the early withdrawal of the Australian helicopter units, which they rated very highly indeed.

An early major operation as the war increased in intensity was the joint US and South Vietnamese assault on the A Shau Valley, running eighteen miles north and south to the west of the town of Heré and which had been held for some time by Vietcong units by April 1968. Described by the US High Command at the outset of the operation, code-named 'Delaware', as a "reconnaissance in force", the main objective was to displace the Vietcong and ensure that they could not use the road running through the valley floor to move their supplies. The operation began on 19th April, when the US Army started to position artillery with the aid of helicopters, including the new Sikorsky S–64 flying cranes and the giant Sikorsky S–65, which moved artillery pieces and ammunition to the top of the hills which surrounded the valley and also rose from the valley floor. This far all went well, but on the second day, events took a turn for the worse, to the extent that newspaper reports were to describe the operation as having "suffered a disastrous start", with some twenty-seven American helicopters being lost on 20th April, that being the highest daily loss rate during the war so far. In a situation which rapidly became chaotic, several helicopters were lost due to collision in the dust raised by their rotor blades at some of the landing sites, while elsewhere, pilots could only land in jungle clearings made by 'daisy cutter' 3,000-lb bombs. Even CASEVAC helicopters had to hover

124

A USAF Sikorsky CH–3C brings some much-needed additional artillery support to a Vietnam battlefield.

A USMC Boeing-Vertol CH–46 Sea Knight at Nui Dang during the Vietnam War.

above small clearings to winch the wounded to safety.

One Sikorsky S–64 flying crane was brought down by Vietcong gunfire from 6,000 feet, although the valley floor was some 2,500 feet above sea-level and the sides rose abruptly to 5,000 feet. A large Boeing Chinook was shot down through clouds. During the campaign, the Vietcong used one of their latest tricks, consisting of a terrorist lying in a bomb crater, which acted as a sound reflector, then shooting straight up into the air as the sound increased and reached its peak, but it also appeared that the Vietcong's anti-aircraft gunners were using some simple form of radar, based on a revolving scanner, which was detected by American helicopter pilots as interference on their radio communications, occurring every twenty seconds or so. As the operation progressed, the United States imposed an embargo on news about it, although by the end of the month the Americans were able to claim success, but at high cost in aircraft and men.

There was, however, a degree of overkill in certain American operations in South Vietnam, using both helicopter gunships, the name which evolved during the war for heavily armed helicopters, and fighter aircraft to flatten villages, such as Phuoc Yen, infiltrated by the Vietcong, or, on 2nd June 1968, using helicopter gunships armed with air-to-surface unguided rockets to flush out snipers in Cholon, the Chinese quarter of Saigon.

Most of the helicopter operations during the Vietnam War were by the United States and her allies, but a few of the Mil Mi–4 helicopters operated by regular North Vietnamese forces did make an occasional appearance in the demilitarized zone, with some being shot down by the American patrols in the area. There was some limited use of helicopters by North Vietnam to supply Vietcong units operating in remote areas of the South, but this was never extensive.

The South Vietnamese forces, on the other hand, operated large numbers of Bell 204 and 205 Iroquois helicopters. Generally, these were used in the same way as the US helicopters, although with a greater emphasis on transport and assault rather than the attack role. Often, helicopters provided the most reliable and potentially most successful means of reinforcing or resupplying beleaguered South Vietnamese government forces. A typical operation was the movement of South Vietnamese troops by helicopter to the embattled town of Kompang Thom, eighty miles north of Phnom Penh, which had been surrounded by Vietcong forces for five days.

It was during the Vietnam War that the concept of using helicopters to rescue air-crew shot down behind enemy lines finally evolved, in spite of an earlier isolated instance of this type of operation at Suez.

The longer-range combat rescue missions of Vietnam brought the need for in-flight refuelling of helicopters from KC–130 Hercules, an HH–3 Jolly Green Giant tops up.

One of the most ambitious rescue tasks, an attempt to rescue large numbers of American prisoners of war from their camp in North Vietnam, however did not succeed. On the night of 21–22nd November 1970, a large force of American helicopters was sent to a POW camp just twenty miles from the northern capital of Hanoi, under the command of Brigadier-General Leroy Manor. On arrival at San Tey camp, the force discovered that the camp, with prisoners, had been evacuated, and indeed appeared to have been deserted for weeks. In spite of the North Vietnamese having fired some thirty surface-to-air missiles at the helicopters, there were no casualties although, in the United States, the exercise was reported as a fiasco. The raid had been prompted by reports that American prisoners were dying in the camps, and had been preceded by a pathfinding mission by USN fighter-

The sequence of an in-flight refuelling session, still something of a rarity for helicopters.

bombers, which had illuminated the area with flares. The helicopter crews on the mission had been lucky, for a little over four years later, a large troop-carrying Boeing-Vertol CH–47 Chinook was shot down by a heat-seeking SA–7 Grail man-portable surface-to-air missile, able to shoot down helicopters while flying below 8,000 feet. On this occasion, there were fifty troops on board; all were killed.

In spite of the seeming difficulty, or even impossibility, of their mission succeeding, not all combat air-rescue operations were failures, indeed, the record of success certainly justified the operation and the commitment of special units to this exacting task. The first success had been on 27th July 1965, when Captain George Martin, USAF, flew to a position just thirty-five miles from the North Vietnamese capital of Hanoi, and during the first five years of the war, well over a thousand American and Vietnamese air-crew were rescued by the Aerospace Rescue and Recovery Service's HH–3 Jolly Green Giant helicopters. It was during the Vietnam War that the concept of in-flight refuelling for helicopters evolved, enabling the helicopter, and in particular those of the rescue units, to remain on patrol for extended periods. A fleet of Lockheed C–130 Hercules transport aircraft was specially modified to act as flying tankers for the helicopter force, becoming KC–130s in the process, because the standard USAF tanker aircraft, the KC–135, a member of the Boeing 707 family, could not fly safely at the speeds necessary to refuel helicopters, while the turbo-prop KC–130 was also a more suitable aircraft for low-altitude refuelling. The concept of in-flight refuelling for helicopters had been attempted during the earliest days of experiment, using road vehicles and a de Havilland Canada Otter aircraft, while helicopters could also refuel from warships without landing. However, it was the Vietnam War which showed this technique to best advantage.

The toll of the Vietnam War on the US forces was heavy. On 16th September 1970, the Vietcong destroyed seven American helicopters and damaged eight others, killing four servicemen and injuring six. On the 4th January 1972, they shot down four helicopters in fighting around the capital of South Vietnam, Saigon and, in this case three of the stricken helicopters were casualty-evacuation machines, although no one was killed and all four helicopters were recovered later. By this time, the Americans had taken to painting their casualty-evacuation machines white, in the hope that the Vietcong might respect their humanitarian mission: it was a vain hope! These daily tolls were not exceptional.

Yet, there were successes, and often well deserved in a war which, if it was not always fought competently or with sufficient regard for

American troops prepare to board a giant Boeing-Vertol CH–47D Chinook in Vietnam.

The Vietnam War posed the need for leaner faster armed helicopters, such as this Bell AH–1 HueyCobra, diving over the jungle.

innocent lives, was nevertheless a necessary war fought for the right reasons. Typical of the successes, and reminiscent of some of the Korean War operations, was the airlifting into Le Minh of a platoon of thirty South Vietnamese troops on 14th January 1974, with cover from helicopter gunships and fighter-bombers. Le Minh, in the West Pleiku province, had been lost to the Vietcong on 22nd September 1973, but was recovered the following January with little fighting.

Some idea of the scale of the fighting can be obtained by the fact that, by the second anniversary of the cease-fire on 15th June 1974, there was a marked increase in fighting, although at the start of the cease-fire, violations had fallen to just sixty a day!

The ferocity of the fighting and the vulnerability of the helicopter while flying troops into a confined jungle clearing surrounded by hostile forces, led to greater use of gunships, which would strafe the surrounding area with machine-gun fire or launch a rocket attack, in effect providing covering fire for the assault force's helicopters. At first, these helicopters were versions of the UH–1, but later, attack helicopters were developed, initially evolving out of the UH–1C to form the Bell Model 209 or AH–1 HueyCobra. The narrower fuselages of these machines, the lower frontal profile which made them a more difficult target and the reduced airframe weight which increased their warload, all helped to give an edge over the traditional helicopter. At first, these machines provided protection for the UH–1 series and for the larger Boeing-Vertol CH–47 Chinook and Sikorsky CH–53, but later this type of machine was to establish a definite niche of its own, which we will look at more carefully in the next chapter.

Two smaller helicopters which also saw service in Vietnam, to some extent filling the gap left by the growth in size of the UH–1 series as the 205 variant became the definitive production machine rather than the smaller 204, were the Bell OH–58A Kiowa and the Hughes OH–6 Cayuse.

After an inauspicious start, failing a United States Army design competition for a light observation helicopter, or LOH, in 1962, Bell continued work on the design as its Model 206, intended to reinforce the company's position in the civil market. This, one of the most successful civil helicopters yet, made its first flight at Fort Worth in January 1966, although in its civilian form, the cabin and fuselage differed rather from that originally proposed to the military. However, the US Army reopened the design competition in 1968, and on this occasion the Bell 206, known to civilian customers as the JetRanger, was selected, retaining its civilian fuselage and cabin layout, as the US Army's OH–58A Kiowa, while later the USN also selected a version of

One helicopter which also appeared in military guise during the Vietnam War was the Bell OH–58A Kiowa, but this is an Austrian version.

Another light helicopter of the period was the Hughes OH–6A Cayuse, and this is one in Colombian Air Force service.

the helicopter as a primary helicopter training aircraft, designated the TH–57A SeaRanger. More than 2,000 Kiowas were built for the US Army, most of them for service in Vietnam, where they were used for observation, tactical reconnaissance, communications and liaison duties, while others were built for the armed forces of Australia, Brazil and Canada, while Agusta in Italy produced this machine alongside the other Bell designs. The Australian Army machines were assembled in Australia by the Commonwealth Aircraft Corporation. A number of Kiowas found themselves in a more active combat role than originally intended, carrying rocket and grenade launchers, complementing the efforts of the larger helicopters.

Not surprisingly for a helicopter with a very similar specification, and with a civilian counterpart aimed at the same segment of the commercial market, the Hughes OH–6A Cayuse was designed for the same US Army competition which inspired the Bell OH–58A Kiowa, originating as the Hughes 369, but winning the first round of the design competition. After a first flight on 27th February 1963, some 1,500 machines, designated OH–6A, were ordered by the United States Army for observation and other related duties, including light transport, with up to four or five troops being carried in addition to the crew of one or two, and also for reconnaissance and CASEVAC duties. The OH–6A was also used to carry XM–27 grenade launchers, pods of unguided air-to-surface rockets or machine-guns. A civilian version, the Hughes 500, was also developed, using the same basic Allison T63–A–5A turbo-shaft as the OH–6A, but de-rated to 280 shp. The Cayuse showed considerable promise from the start, establishing a number of records for helicopters in March and April 1966, including an endurance record of 2,213 miles in a straight line, and a speed record of 171.8 mph over a closed circuit.

Subsequent development of these machines has led to improvements in their performance, while that of the Hughes model in particular has included work on developing quieter helicopters, mainly for the battlefield role and, in recent years, both types of helicopter have also been armed with TOW anti-tank missiles. Like its competitor, the Hughes OH–6A is also built under licence in Italy, by BredaNardi.

During its closing stages, the Vietnam War went from bad to worse. Apart from the lack of observance of the cease-fire, an agreement that neither the North Vietnamese nor the South Vietnamese should receive arms to replace combat losses was observed rigorously by the United States after the withdrawal of more US personnel, but not by the Vietcong's North Vietnamese and Russian allies. Steadily, the

133

balance tipped against the south, and the Vietcong advanced, taking first one town and then another, before finally surrounding Saigon itself. By April 1975, the Republic of South Vietnam was clearly all but defeated, and on 20th April, nine aircraft-carriers and amphibious assault ships attached to the United States 7th Fleet converged on South Vietnam ready for the inevitable evacuation; with many of the carriers, almost equivalent in numbers to the usual peacetime operational carrier force of the United States Navy, equipped solely with helicopters with which to conduct an evacuation under fire. Two days later, 7,000 United States Marines were landed to protect American citizens and also the 13,000 South Vietnamese cleared for immigration into the United States, while clearance procedures for many more were accelerated. After a week, the evacuation was in full swing, in an atmosphere of growing chaos and panic.

On 29th April, the command ship, USS *Blue Ridge*, handled fourteen helicopters in a period of just ninety minutes on a landing platform intended for just one helicopter. The passengers amongst this influx included, amongst others, Air Vice-Marshal Nguyen Cao Ky, former Vice-President of South Vietnam. However, the statistics, impressive though they may be, conceal the real panic and urgency of those ninety minutes, during which the real miracle was simply that no one was killed. The first South Vietnamese helicopters arrived over the American fleet while the arrangements for their reception were incomplete, and the first of a group of seven helicopters to land aboard the USS *Blue Ridge* hit another helicopter on the landing-platform ready to take off. The South Vietnamese pilot dropped his machine directly onto the whirling rotor blades of the helicopter already aboard the ship, scattering fragments of metal across the landing pad and with the new arrival nearly crashing into the sea, before the doors opened to drop screaming women and children onto the deck. Later during that same sad morning, another South Vietnamese helicopter pilot jumped from his machine before it landed, and the helicopter then crashed onto the deck as a consequence of this premature abandonment, and again the whole incident became rather messy, with pieces of whirling metal scattered across the deck.

A procedure was established: South Vietnamese helicopters landed, passengers were hastily unloaded, doors were ripped off so that the helicopter could sink quickly and so that the crew could escape, and then the machine would take off and land in the sea, from whence its crew, often just the pilot, could be rescued by waiting American helicopters. On the larger ships, unloaded helicopters were simply pushed overboard, minus the crew, of course! However, at the end of

134

Back to Vietnam, and away from it. A South Vietnamese Iroquois is flown
into the sea during the evacuation from Vietnam, shortly before the Vietcong
overran Saigon.

the operation, a large number of South Vietnamese helicopters were
brought away from the area aboard the American ships. Although
these machines were in any case American property, having been given
to the South Vietnamese through American military aid, this did not
prevent the Vietcong from claiming their return.

The war was a tragic farce. United States policy was subsequently
proved to be correct as first South Vietnam and then Cambodia fell,
and their populations bore the brunt of a particularly severe and
fundamentalist brand of Communism. Possibly, many of the methods
used by the Americans were too heavy-handed, something illustrated
by such statistics as the dropping of 6.75 million tons of ordnance by
the US forces, compared with a combined British and American World
War II total of just 2.7 million tons. No less than 8,000 American

aircraft were lost in the war, and of these, some 4,600 were helicopters of one kind or another.

While today many counter-insurgency operations are conducted by governments with the aid of their allies, possibly reflecting the increased role of outsiders in fomenting and supporting insurgency, there are still cases in which the resident government has to stand alone. The prime, indeed, the most extreme example of this was the civil war in Rhodesia, where a United Nations embargo on any dealings or relations with the Rhodesian Government, or even with companies and individuals resident inside Rhodesia, prevented assistance with the exception of some relatively limited military aid from neighbouring South Africa.

At the time of Rhodesia's so-called UDI, the unilateral declaration of independence on 11th November 1965, a blockade was imposed by the Royal Navy in the Mozambique Strait to prevent oil for Rhodesia from being shipped through the port of Beira in Mozambique, at that time officially part of Portugal. The so-called "Beira Blockade" was a futile operation, as well as an unpleasant and boring task for the Royal Navy, absorbing the services of one major aircraft-carrier, either HMS *Eagle* or her sister ship, HMS *Ark Royal*. It did not provide any extensive workload for the Royal Navy's helicopters.

Ashore, land-locked Rhodesia's armed forces consisted of a small

The French also build good COIN helicopters, such as this Alouette II, in ALAT (French Army) service.

mainly conscript army and a small air force, with a single squadron of each of Hawker Hunters and English Electric Canberra jet bombers, a mixed counter-insurgency and training squadron with Percival Provosts, and a mainly Douglas C–47 transport squadron, while a small helicopter force was equipped with French-built Alouette II and IIIs. This was a small but capable force for a country with just 200,000 white settlers providing all of the finance and most of the manpower for the armed forces at the time of UDI, provoked by a serious disagreement with the British Government over the future of the country and its independence. Curiously enough, the stresses and strains of the guerilla war which followed UDI forced integration on the armed forces.

However, at first there was little or no warfare of any kind. Royal Air Force Gloster Javelin jet fighters based in Zambia at Lusaka Airport to protect that country from an imagined threat from Rhodesia tended to use Salisbury air-traffic control in preference to that of the host country! It would have been hard to imagine any type of conflict between the British forces and those of Rhodesia, modelled on the British pattern and with strong bonds between the two nations, while Rhodesia had been the World War II training base for many British RAF air-crew; Rhodesia had been a major element in the wartime Commonwealth Air-crew Training Scheme. Gradually, however, terrorist activity started to develop, with at least the covert if not the overt approval and support of the governments of neighbouring African states, several of whom provided bases for one or the other of the two main Rhodesian terrorist groups. This activity intensified with the Portuguese withdrawal from their African territories, Mozambique and Angola, where terrorist activity was rife for some time before independence. The helicopter did not play any real part in the wars in the Portuguese territories, partly because of the limited number of helicopters in Portuguese service, due to the relative poverty of the country and also because of the reluctance of many countries to sell Portugal military equipment at the time, in spite of her membership of NATO.

By the early 1970s, the situation in Rhodesia had changed for the worse, with clashes between Government troops and the guerillas on a daily basis, and farms and other settlements having to be defended by their occupants while able-bodied whites of military age spent up to half of their time on active service. Roads were mined extensively, and on many routes travel was only possible by convoy, which meant that in this country of fairly sparse population, the helicopter became the main means of rapid reinforcement. Yet, the embargo on arms or even consumer goods to Rhodesia meant that the only helicopters available were the Alouette IIIs, even though the larger Boeing and Sikorsky helicopters used in the Vietnam War would have been a godsend to the Rhodesian troops, as indeed would have been an obsolescent type such as a Wessex or even a Whirlwind. This meant that Rhodesia's helicopters had to be used for "hot pursuit" or "hunt and kill" operations, with the fleet of ageing Dakotas also dropping small groups of paratroopers in an attempt to improve the Rhodesian Army's ability to reach troublespots quickly enough to be effective.

The Alouette III was, as the name suggests, a development of the highly successful Alouette II helicopter. Originating as the SE.3160, the Alouette II was redesignated the SA.316B on the formation of Aerospatiale, which subsequently became the aircraft's manufacturer. Powered by a 570-shp Turbomeca Artouste IIID turbo-shaft, and offering a larger cabin, the six-passenger Alouette III made its first flight early in 1959, with the first production machines being delivered just three years later. The production SA.316B and the more powerful SA.316C, with an up-rated 600-shp engine, entered service with a large number of customers, mainly military, including Rhodesia and many countries in the third world, more than 1,500 being built. The French armed forces proved to be the largest customers, with the

An Alouette III in ALAT service, a larger machine than the earlier II. Note the rocket pods.

Army and Air Force each taking about a hundred machines, while the Aeronavale took just over twenty, and the helicopter was also built under licence in India by Hindustan Aircraft, as well as being assembled in Romania and Switzerland. Although a light helicopter, the Alouette III could carry two passengers and two stretchers as well as a single crew member instead of the normal load of up to six occupants, or a slung load of 1,600 lbs, about three-quarters of a ton, could be lifted. Although not equipped with a dunking sonar, the helicopter was used for reconnaissance at sea, and could support anti-submarine operations flying in the killer role carrying torpedoes.

Perhaps the true worth of the Alouette III was discovered in Rhodesia, where the difficult conditions meant that the helicopter frequently went into battle carrying an overload, necessitating recourse to the practice of a running take-off, effectively demoting the helicopter to the status of a STOL, short take-off and landing, machine, in order to obtain the necessary lift.

Typical of the uses to which the Rhodesian helicopters were put was the killing on 1st October 1974, of two guerillas who had been responsible for the murder of five people in the Chivreshe tribal trust land during the preceding ten days. After an intensive search by the Rhodesian security forces, the guerillas were spotted and shot dead from the helicopter after they opened fire on the Alouette III with their Russian assault rifles. Such were the pressures on the Rhodesian security forces that, when the Portuguese settlers in Angola appealed for help in August 1975, Rhodesia could do little, although South Africa dispatched a token force to conduct a limited operation with some helicopter support, until forced by pressure from the United States to withdraw.

In spite of the excellent service given by the Alouettes, there were the inevitable casualties. Major-General John Shaw, Rhodesian Army Chief-of-Staff, was amongst several officers killed in a helicopter crash, near Umtali on the Rhodesian border with Mozambique, after the helicopter struck a high tension overhead power cable just before Christmas 1975. This was a rare cause of a helicopter accident in Africa, although it is an all too common occurrence in the developed nations, including the British Isles.

Eventually a solution to the Rhodesian problem was found which was acceptable to the British and American governments, the governments of the neighbouring African countries, the different warring guerilla groups within Rhodesia itself, and a substantial proportion of the white population. This led to one final role for the helicopter in Rhodesia, collecting and delivering ballot-boxes and electoral officials to and from the different polling stations throughout the country in its first majority rule general election during April 1979. Mercifully, the election led to an almost complete end to guerilla activity, although not at first its complete elimination. The Rhodesian crisis was over! During the period before and during the election, a considerable number of British helicopters, mainly Army Air Corps' Lynx and Gazelle machines, with a few Royal Air Force Pumas and some Royal Navy helicopters, were sent to Rhodesia, to assist in rounding up the various guerilla bands and resettling them in camps under British control, and then later to play their vital role in the election itself. Sadly, having withdrawn and then sold the Royal Air Force's giant Short Belfast transport aircraft as an economy measure, the larger Royal Air Force and Royal Navy machines had to be transported to Rhodesia aboard United States Air Force transport aircraft, denoting a decisive weakening of British capability, not least in the ability of politicians to plan ahead for the unexpected or the essential. So much for Bersatu Pardu! There were no major individual operational incidents affecting the British helicopters during their stay in Rhodesia, since their use was confined to the transport, communication and liaison roles.

While the Americans were still tied down in South Vietnam and the Rhodesian crisis rumbled on, the British had a major counter-insurgency problem of their own which started not long after Rhodesia's unilateral declaration of independence, and continues to this day. From 1969 onwards, there has been growing terrorist activity within the United Kingdom itself, in Northern Ireland, as the Communist-backed Irish Republican Army, the IRA, and the even more extreme Provisional IRA, with such militant rival but similarly motivated minor terrorist groups as the Irish Liberation Army, embarked on a campaign of political pressure and shootings and bombings aimed initially at forcing a British withdrawal from Northern Ireland and then at installing a revolutionary left-wing government in what they hoped would be a united Ireland after the withdrawal. Northern Ireland covers six of the nine counties of the old Irish kingdom of Ulster, and remained separate from the other twenty-six Irish counties in 1922 when the British Government offered independence to Ireland. As civil wars go, the trouble in Northern Ireland has been comparatively steady, with the IRA campaign worsening and then slackening, and occasional acts of terrorism from extreme anti-Irish republican groups contributing to the overall state of chaos, but the death toll, at some 2,000 people during the first twelve years, is far less than in other conflicts elsewhere in the world.

140

The helicopter could not remain out of the Northern Ireland crisis for long. The rolling green hills of the northern part of Ireland and the often heavily indented coastline provided a different scenario from the jungles of south-east Asia and the bush of Rhodesia, or indeed the desert of Aden. However, this remained good terrorist country and good helicopter country, with the sometimes heavy mining of roads and railway lines further encouraging the use of the helicopter by the security forces. Initially, the British Army used Bell Sioux and Westland Scout helicopters, while the Royal Air Force, for larger-scale troop movements, used the Westland Wessex, but gradually this force was updated, with the substitution of the Westland-Aerospatiale Gazelle and the Westland Lynx in Army Air Corps service, and the Westland-Aerospatiale Puma superseding the Wessex of the Royal Air Force. The Puma, of which more later in the next chapter, could move up to twenty troops at speeds of up to 150 mph, offering some considerable scope for rapid intervention.

Initially, helicopters were used mainly on communications and liaison duties, although British troops had practised firing machine-guns from Sioux helicopters during trials on the firing-range at Magilligan Point, in County Londonderry, as early as 5th April 1971, marking the start of helicopter operations in Northern Ireland. However, the Vietnam and Rhodesian concepts of the helicopter as a gunship have not been applied by the British forces in Northern Ireland.

One of the first extensions of the helicopter's role in Northern Ireland was its use for close reconnaissance after mining incidents to look for the tell-tale wires which warned of a booby-trap for troops and police moving into the area. Often, the booby-traps would also be applied to railway or road vehicles hijacked by the IRA or other left-wing militant groups. This type of extremely close reconnaissance rapidly became a frequent duty for the Army's helicopters. Border patrols were also flown, and some attempt was made at "hot pursuit" of terrorists, but in spite of pressure by the British Government on its Irish counterpart, helicopters were not allowed to follow terrorists over the border into the Irish Republic, providing the terrorists with ample scope for sudden raids into the border areas, with a hasty withdrawal back into the Republic. Helicopters were also used for monitoring demonstrations and protest marches by IRA sympathizers, usually in Belfast or Londonderry. Bomb-disposal experts in Northern Ireland frequently arrive by helicopter, as the quickest and safest way of moving these specialists to wherever they may be needed.

Essentially, the type of operation conducted by the military in Northern Ireland is far lower key than that in Vietnam or Rhodesia, attempting to allow as much of the population to lead a normal life as might be possible, and giving the IRA few excuses for the level of violence in the province. While the helicopter is a key element in military aviation in the area, it does not have the support of other more glamorous combat aircraft, as in Rhodesia and Vietnam, while, as already mentioned, there are no gunship operations even by the helicopters. Perhaps the key to this different approach is the overwhelming support for the union with the rest of the United Kingdom to be found amongst the province's civilian population.

However, even in Northern Ireland the helicopter has had its moments, and one of the most significant operations for the helicopter in Northern Ireland came during 1971. After a rising level of violence, the then Government of Northern Ireland (a body which no longer exists and was in any case a subordinate regional government to Westminster), decided to impose internment, or imprisonment without trial, for those known to be active members of the IRA and other terrorist organizations. The move was the answer to a number of gun battles in the major cities of Northern Ireland and to the intimidation of witnesses, a familiar IRA tactic, which made normal prosecution for terrorist offences difficult. On 9th April, internment was introduced under the Special Powers Act, with troops and members of the Royal Ulster Constabulary being carried in Army and Royal Air Force helicopters in an operation starting at 04.30 that morning, acting quickly and successfully to round up about 70 per cent of wanted persons, some three hundred in all. Many Northern Ireland politicians both at the Northern Ireland Parliament at Stormont and at the United Kingdom Parliament at Westminster, however, criticized the operation for being too late, although this reflected on the introduction of internment as a measure rather than on the ability of the security forces, or their helicopters to act quickly once given the go-ahead.

Not all of the operations in Northern Ireland were without incident for the helicopter itself. At several times there have been fears that the IRA had in its possession Russian-built SA–7 Grail man-portable surface-to-air missiles, which would be devastatingly effective against helicopters. On other occasions, helicopters, mainly those of the Army, have come under fire, especially in the Crossmaglen area, almost on the border, with the IRA's favourite weapon being the M60 machine-gun. The early loss of an Army helicopter in 1973 while flying between Gough Barracks in Armagh and Aldergrove Airport, the

Royal Air Force's main base in the province, was an accident. However, on 17th February 1978, an Army Air Corps Gazelle crashed near Newry, South Armagh, after being shot at, killing all aboard, including the passenger, Lieutenant-Colonel Ian Corden-Lloyd, the commanding officer of the 2nd Battalion, The Royal Green Jackets, an English light infantry regiment.

Not all the helicopters used against the IRA were British. The threat to the security and stability of the Irish Republic posed by the IRA forced that country to take defence seriously for once (it is the only EEC member which is not also a member of NATO), and in addition to some co-operation between the British and Irish armies, depending on the political climate between London and Dublin at any time, the Irish Army has intensified its own border patrols and also been allowed to expand and re-equip. Irish Army Air Corps Aerospatiale Alouette III helicopters conduct border patrols, and are also used to move the two bomb-disposal squads, based close to the border at Sinna, County Donegal and at Dundalk, to wherever they may be needed, while additional bomb-disposal squads are based at Dublin, Cork, Athlone and at the Curragh. An indication of the level of activity south of the border is that during the peak twelve months from September 1973 to September 1974, there were some six hundred incidents requiring the assistance of the Irish Army bomb-disposal squads, although many of these would have been false alarms.

Partly as a result of their experiences in Northern Ireland, a joint Metropolitan Police and British Army security exercise was mounted at London's main airport, Heathrow, in October 1974, with helicopters being used to assist in troop movement and in surveillance.

In less than its own lifespan, to date, indeed, over a period of some thirty years, the helicopter has become the primary counter-insurgency weapon, giving security forces hitherto unknown flexibility and versatility in opposing the growing threat to world stability and even to peace itself posed by the growing number of revolutionary organizations in the world today. Indeed, as one Russian dissident, Alexander Solzhenitsyn, suggests that this obviously related and co-ordinated level of violence is indeed World War III, the prelude to a global World War IV, and that the West is losing the battle, then the helicopter has obviously an active career still ahead of it. There are too, times when it is difficult to assess the difference between battlefield and counter-insurgency duties, as in Afghanistan, for example, and we look at this aspect of the helicopter's development in the next chapter.

143

7

ON THE BATTLEFIELD

As first conceived, the helicopter was considered as having a relatively limited battlefield role, centred mainly around aerial observation post and liaison duties. The experience gained during the Korean War showed just how important the helicopter could be for casualty-evacuation and transport duties. Indeed, it would be fair to say that at a period of increasing battlefield mobility, the helicopter was able to make a major contribution, providing even greater mobility for troops and, later, heavy equipment, and offering the possibility of encircling an enemy position, the concept of "vertical envelopment". The worse the terrain, the greater the advantage bestowed by the helicopter and the lower the helicopter's vulnerability to counter-attack or defensive fire. Both the Korean War and the Suez operation could be regarded as early examples of the battlefield mobility and the other applications of the helicopter, although the growing demand for helicopters in the counter-insurgency role has raised the paradox that guerilla warfare tends to be the battlefield of the late twentieth century.

The role of the helicopter has expanded in this area as in others, and few of the strong advocates of the helicopter on the battlefield could have foreseen not only the growing importance of the helicopter for ferrying troops at high speed across difficult terrain to new positions, but also the other tasks which have befallen it, including the development of the anti-tank helicopter and, perhaps more predictable, the flying crane. It was in Vietnam that the flying-crane helicopter first showed its paces, its potential for bridging, heavy supply and even the rescue and recovery of other helicopters and military aircraft after they had crash-landed or been damaged on the ground. Few too could have foreseen military commanders having sufficient confidence in these frail machines to risk sending them behind enemy lines on rescue missions, although this was a logical extension of the helicopter's assault role.

Possibly, the biggest boost to the battlefield support role of the helicopter came with the advent of vertical take-off combat aircraft, of which the first true example was the Hawker Siddeley, later British Aerospace, Harrier; the AV–8A in United States Marine Corps service. An aeroplane able to operate away from fixed bases without runways, indeed even in theory at least able to operate from behind

One of the first helicopters designed specifically for Army use was the Saunders-Roe Skeeter, seen here with the British Army.

A Westland Scout fires an AS.11 anti-tank guided missile.

enemy lines, requires a supply system with similar capabilities, able to ignore rough terrain, to be independent of roads, bridges or railways, and of course, also able to dispense with runways, for, after all, the helicopter was *the* original vertical take-off aircraft. Unfortunately, severe budget restrictions prevented the Royal Air Force from operating effectively in this way at first, with its Belvedere medium-lift helicopters retired in 1969 and an order for an American machine cancelled shortly afterwards, and indeed not reinstated for some years afterwards so that the RAF first put medium-lift helicopters, Boeing-Vertol CH–47 Chinooks, into service in 1981.

Some of the helicopters used on the battlefield evolved with this application, or range of applications, in mind, while others were developed for other roles and came to the battlefield almost reluctantly. One helicopter design which was developed with both small-ship operation at sea and battlefield duties very much in mind was the Westland Wasp and Scout series. While their designs were about 75 per cent common, these helicopters included some important differences, with the Wasp being the naval version and the Scout being the Army machine. It was, taken as one project, that relatively rare item, a highly successful British helicopter.

The Scout originated as a Saunders-Roe project for a Skeeter replacement, the P.531, with two prototypes making their first flights during July and September 1958, and both used a 325-shp Blackburn-built Turbomeca Turmo turbo-shaft. The following year, Westland Aircraft acquired Saunders-Roe and continued development of the P.531, but changing the course of the development by using a more powerful 635-shp Blackburn Nimbus turbo-shaft on the third prototype and an even more powerful Bristol Siddeley Gnome 685-shp turbo-shaft on the fourth prototype, which flew for the first time on 3rd May 1960.

By this time, the Scout had also developed from being a purely private venture exercise into one with the support of the British Army, which had ordered a pre-production batch of Scouts in 1959, and after tests with the first of these machines in 1960, the first of a number of production orders was placed for the Army Air Corps. Initially, the Scout replaced the Army's Skeeters, but the utility of this five-seat Rolls-Royce Nimbus 101 685-shp turbo-shaft-powered helicopter was such that it also played an important role in the virtual replacement of fixed-wing aircraft in the British Army. In addition to being used for liaison, communications, AOP and CASEVAC operations, carrying two stretchers inside the cabin and two externally, the Scout was used first for experiments and then with operational squadrons for anti-tank operations, using the Aerospatiale AS.11 wire-guided anti-tank missiles.

A contemporary of the Scout, the Sud, later Aerospatiale, SA.321 Super Frelon, also made its first flight in prototype form during 1959, but with two differences; the first being that the Super Frelon was a far larger helicopter, while the machine was designed initially for naval use, entering army and air force service later in Israel, Iran, Libya and South Africa. Strangely enough, the Super Frelon has never been operated by the French Armée de l'Air or the ALAT, French Army Aviation, both of which restrict themselves to the smaller Puma, rather than utilizing the Super Frelon's considerable carrying capacity of up to thirty-seven passengers at speeds of up to 155 mph. Unusually, at the time, the Super Frelon was developed as a single-rotor triple-engined helicopter, incorporating considerable assistance in the design of the rotors from Sikorsky in the United States, but using French Turbomeca Turmo turbo-shafts with independent drive from each engine to the rotor, providing superb engine-out capability.

The original prototype of the Frelon, or Hornet, was designated the SA.320, and the first of two such machines flew on 10th June 1959, having been designed to meet the needs of all three French armed forces. In prototype form, the Frelon used Turmo turbo-shafts of just 800 shp each, and both the rotor blades and the rear fuselage folded to aid storage, especially aboard warships. However, further development led to the SA.321 Super Frelon prototype, with watertight boat-form hulls and rear-loading ramps for troops, cargo or small vehicles rather than a folding rear-fuselage, although the extreme tip of the tailplane could still be swung out of the way, and it was in this form that the Super Frelon's first flight took place on 7th December 1962. A second prototype incorporated search radar in the stabilizing floats on either side of the fuselage, and both prototypes used far more powerful 1,320-shp variants of the Turmo engine, the IIIC2: indeed, the improvement in performance was quite remarkable, with one of the prototypes establishing several speed and altitude records during July 1963, including one for straight line speed of 217 mph, or 350 kmph, which remained unbeaten until 1970, and which was some 60–70 mph higher than the machine's true cruising speed. After flights with four pre-production machines, the first of which flew in November 1965, deliveries of production SA.321Gs to the Aeronavale started in 1966, with the machines operating from the helicopter carrier *Jeanne d'Arc* and on anti-submarine patrols from shore bases, and also possessing the capability for conversion for minesweeping. While the Aeronavale machines carry anti-submarine torpedoes or air-

147

to-ship missiles, such as Exocet, the helicopters in service with the Israel Defence Force, the South Africa Air Force, the Libyan Air Force and the Iranian armed forces are all used for troop movement and supply duties, without even the unguided rockets sometimes fitted to assault helicopters. A small number of Super Frelons have also been supplied to the armed forces of the Chinese People's Republic, although production in total is only slightly above the one hundred mark.

The French concept of a medium lift helicopter, the Aerospatiale Super Frelon.

In tight formation, three Westland–Aerospatiale Puma helicopters of the Royal Air Force carry troops to their destination.

From the same stable, the smaller, but markedly more successful, Aerospatiale SA.330 Puma, was developed by Sud Aviation to meet a French Army requirement for a medium-sized assault and transport helicopter capable of all-weather operation. The first of two prototypes made its first flight on 15th April 1965, and was followed by six pre-production machines before the first production Puma was built in September 1968. That same year, agreement was reached between Aerospatiale and Westland Aircraft in Britain to produce the Puma, the Gazelle and a Westland design, the Lynx, as a joint helicopter programme, entailing Westland assembly of some forty Pumas for the Royal Air Force and construction of components and assemblies for Pumas built in France for use by the French Army and Air Force and for export to third-world countries. Production military Pumas used either twin 1,400-shp Turbomeca Turmo IVB or 1,575-shp Turbomeca Turmo IVC turbo-shafts, giving a maximum cruising speed of some 170 mph, with up to twenty troops being carried, or six stretchers and six seated persons, in addition to a crew of two.

Almost 150 Pumas entered French Army service, while the Armée de l'Air operates a small number of VIP, communications and transport duties, and in addition to the Royal Air Force, Pumas are also operated by Algeria, Abu Dhabi, Belgium, Chile, the Ivory Coast, Portugal, South Africa and Zaire. The helicopter can be used with 20-mm cannon, 7.62-mm machine-guns and either air-to-surface missiles such as Exocet or the AS.11, or with unguided rockets. During the early 1980s, production of the improved Super Puma started, with initial deliveries to the French Army. Although smaller than the Sikorsky S–61 and Westland Commando, and with restricted headroom inside the passenger cabin, which is behind the cockpit and under the engines in the now standard layout for medium and large-sized helicopters, the Puma does have a good reputation as a sturdy and reliable helicopter, with considerable speed and smoothness in flight.

Part of the same Anglo-French helicopter programme, the SA.341 Gazelle is a far smaller helicopter, carrying up to five persons including a single crew member in the unarmed versions, and flying at speeds of up to 192 mph, while being able to carry machine-guns, rocket pods, or up to six HOT air-to-surface anti-tank guided missiles or four TOW missiles or two AS.12 anti-tank missiles.

The first Gazelle prototype was completed by Aerospatiale and first flew in April 1967, while the second prototype was followed by four pre-production aircraft including one completed by Westland to meet British Army requirements. Production machines use a 600-shp

Turbomeca Astazou turbo-shaft, and almost 160 of these were ordered for the British Army and Royal Marines as the SA.314B, or AH Mk. 1 in Army service. The Royal Navy received thirty SA.341Cs as their HT Mk. 2 and the Royal Air Force received just fourteen SA.341Ds as their HT Mk. 3, with both the naval and air force versions being used as helicopter trainers. The French Army received more than 160 Gazelles, using these in the same communications, liaison and anti-tank roles as their British Army counterparts, while the helicopter has also been exported to a small number of other air arms, including the Kuwait Air Force, and is built under licence in Yugoslavia. An improved and up-rated version of the Gazelle is the SA.342.

A helicopter of similar performance to the Puma, but slightly larger, the Russian Mil Mi–8, known to the NATO alliance as 'Hip', first flew some time during late 1961, with the prototype incorporating a number of components from the earlier and smaller Mi–4, using a single 2,700-shp Soloviev turbo-shaft mounted above the cabin and behind the flight-deck. Since becoming one of the standard Soviet and Eastern Bloc army workhorses, the Mi–8 in its production form uses two 1,500-shp or 1,700-shp Isotov turbo-shafts driving a four-or five-

An attractive lightweight helicopter, the Aerospatiale Gazelle is produced jointly with Westland; this RAF machine is used to train helicopter pilots.

bladed rotor and has a maximum speed of about 140 mph and a capacity of up to twenty-eight troops. Although the basic layout is similar to that of most Western machines, it does appear dated in certain respects, with a non-retractable undercarriage, for example, possibly because this machine is slightly slower than some of its Western counterparts, and the elimination of drag with a retractable undercarriage may have been considered by the designers to be of minimal benefit. After entering service with the Soviet armed forces during the later 1960s, the Mi–8 has also entered service with all of the Warsaw Pact nations, as well as with Egypt, Ethiopia, India, Iraq, Pakistan, Peru, Sudan and Syria, and in small numbers with a few African states and in Vietnam. In addition to carrying troops or casualties, the Mi–8 can also carry an external armament, including machine-guns and rockets.

The Mi–8's partner in many Communist air forces and armies is the Mi–6 'Hook', which was described in chapter 5, and the development of the Mi–6, the Mi–10, and Mi–10K, known to NATO as the 'Harke'. The Mi–10 utilizes the power-plant and rotor system of the Mi–6 and can be considered as a flying-crane development of the earlier machine, although unlike some of the American flying cranes, the Mi–10 does retain a shallow cabin, using the reduced cabin depth and its flat underside, plus a distinctly long-legged undercarriage, to enable the helicopter to fulfil its flying-crane role. The first public appearance of the Mi–10 was in 1961, but in 1966 information became available in the West about a variant, the Mi–10K, with a shorter undercarriage and facilities for carrying slung loads, rather than fitting loads close to the underbelly of the Mi–10 itself. In May 1965, the Soviet Union claimed a record for a Mi–10 of a load of 55,300 lbs being carried to an altitude of 9,320 feet, although operational loads tend to be nearer the 30,000 lbs mark, about 14 tons. In addition to a three-man crew, up to twenty-eight passengers can be accommodated in the cabin of the Mi–10, and operations with a large load are assisted with the use of a closed-circuit television camera, to enable safe landings to be made. On the Mi–10K, a rearward-facing gondola under the fuselage enables a crew member to supervise the lifting and landing of a slung load. In common with most crane helicopters, the Mi–10 and Mi–10K are in service in fairly small numbers, reflecting the usual adequacy of the slung-load capability of standard helicopters.

Largest of the Russian helicopters, the Mil Mi–12 'Homer', is believed to have first flown in 1969, before making its first appearance in the West at the Paris Air Show in 1971. In some ways, the helicopter seems like an outsized development of the Focke-Achgelis Fa.223,

The standard Soviet bloc troop-carrying helicopter is the Mil Mi–8 'Hip'.

Soviet troops exercising in Byelorussia run to their waiting Mil Mi–8s 'Hip' helicopters.

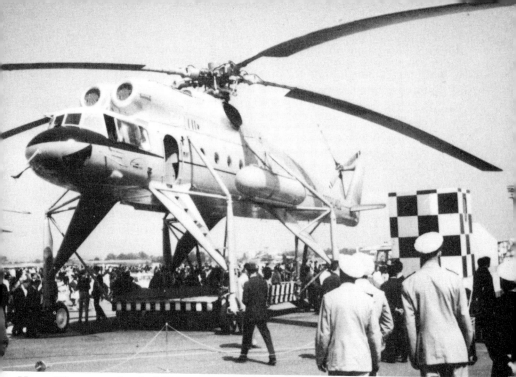

Heavy lift support for the Mil Mi–8 is provided by the Mil Mi–10 'Harke' flying crane.

with large wingtip-mounted rotors side-by-side, each driven by two powerful 6,500-shp Soloviev D–25VF turbo-shafts, giving the helicopter a maximum speed of some 160–180 mph, and accommodation for up to seventy passengers, but the real merit of the machine lies in its lifting capability, which on 6th August 1969, established a world record by taking a load of 88,600 lbs to an altitude of 7,340 feet, that is lifting forty tons, a weight higher than the gross take-off weight of the largest American helicopter. Intended to carry missiles and other items of heavy equipment, the Mi–12 design specification called for a similar payload capability to that of the large Antonov An–22 transport, but it is doubtful whether this has been achieved on test or in practice. A pair of large clam-shell doors at the tail, and a loading ramp, provides access to the huge interior, with an electrically powered crane at the forward end of the fuselage to assist with loading, but an unusual feature is the division of the six-man crew between a nose flight deck and an upper deck cockpit, which accommodates the navigator and radio operator. Few, if any, of these machines have entered service, and it seems that the Soviet armed forces are placing the emphasis on a smaller but more practical Mil helicopter, the Mi–26, for the future. The exact reason for the seeming lack of enthusiasm for the Mi–12 is unknown, but there may be problems in handling, and certainly the first prototype crashed during its test-flight programme in 1969, but it may also be that there are relatively few occasions when such a large helicopter can be used to full advantage, while a smaller helicopter able to fulfil many of its functions offers greater flexibility and the benefit of being available in large numbers.

Large helicopters for military use have also been built in the United States, however, and are almost certainly more reliable and easier to handle than their spectacular Soviet counterparts. One of the most successful of the 1960s generation of American helicopters was the Sikorsky S–61, a twin-turbine, single-rotor helicopter available for military, naval and civilian use, and which has enjoyed considerable success, particularly in naval and commercial use. Although variants of this helicopter have to their credit many of the early combat rescue successes with the United States Air Force in Vietnam, the machine is far better considered as a naval helicopter, and its development and some idea of its service record is dealt with in the following chapter.

As the American manufacturer whose work can be most easily compared with that of the Soviet Mil design bureau, placing the emphasis on single-rotor machines for a variety of civilian, military and naval applications, it is interesting to note that Sikorsky has also been the only American manufacturer to take the concept of the flying-crane helicopter seriously, although Boeing-Vertol has conducted design exercises in the past. Sikorsky originally started work on flying cranes in 1958, building a helicopter designated the S–60 as a design exercise, with the cost being shared between the manufacturer and the United States Navy, which was interested in the concept as a means of improving vertical replenishment of warships under way, and also for the recovery of damaged aircraft. The S–60 utilized the rotor and mechanical components of the large S–56 helicopters, with the engines mounted in the same large pods on either side of the fuselage, although on the S–60 these also acted as the housing for the rectractable undercarriage. A first flight in March 1959 was followed by two years of intensive trials before the prototype was lost in an accident.

However, the S–60 proved the concept's soundness, and work was already well advanced, by the time the prototype was lost, on a prototype S–64, which first flew on 9th May 1962, utilizing a six-bladed main rotor and powered by twin 4,050-shp Pratt and Whitney turbo-shafts. Unlike the Soviet concept of a large flying crane with loads fitted below the fuselage, the S–64 lacks a fuselage or cabin

A Boeing proposal for a heavy lift flying crane and bridging helicopter, dating from the mid-1970s. Development of a new twin rotor machine of this type, the XCH–62, is now in hand.

behind the cockpit, offering instead a boom with more than nine-feet clearance between it and the ground, so that loads of reasonable size, including a standard 8 foot 6 inch container, can be wheeled under the boom ready for lifting, while the crouching landing gear enables the S–64 to squat over its load before lifting it, or even to raise it off the ground to taxi to a take-off point.

The S–64 was the subject of an extensive evaluation by the United States Army before an order was placed for six pre-production examples in June 1963, for delivery during 1964 and 1965 with the Army designation, YCH–54A Tarhe, while a further eighteen followed in 1966 as the production CH–54A. The production machines used Pratt and Whitney T73–P–1 turbo-shafts of 4,500 shp each, although later production models used up-rated engines of 4,620 shp, and the later CH–54B used engines of 4,800 shp each.

Altogether more than sixty Tarhes were introduced by the United States Army, initially operating within the 478th Aviation Company fighting alongside the US Army's 1st Cavalry Division in South Vietnam. Although only capable of a maximum speed of 130 mph, the Tarhe can usually lift a load of some ten tons, or motor vehicles or, using special containers, a field hospital unit, forty-eight stretcher cases or up to sixty-seven troops, although during the Vietnam War there was one recorded instance of eighty-seven troops being lifted. On 12th April 1972, a record was established with the lift of a payload of more than 34,000 lbs to 10,850 feet.

Sikorsky's other large helicopter is the S–65A, which is the West's largest and heaviest helicopter, although it still retains the famous Sikorsky single-rotor system. Originally developed using the S–64 rotor and transmission systems, with a fuselage similar in appearance to that of the S–61, but, of course, considerably larger, the S–65 was developed after winning a 1962 United States Marine Corps competition for a ship-borne heavy assault helicopter, and designated the CH–53A Sea Stallion. A prototype flew on 14th October 1964, followed by the start of deliveries of an initial order for 106 machines in September 1966. Introduction to service went smoothly, and within six months of the first helicopters being received, the CH–53A was serving with the United States Marines in South Vietnam. Powered by twin 2,800-shp General Electric T64–GE–6 turbo-shafts, the CH–53A

Strategic mobility – two USAF HH–3 helicopters are loaded into a C–133 transport during the Vietnam War. The helicopter has often to hitch a ride, even today.

can lift almost forty fully equipped troops, twenty-four stretchers and four medical attendants, or up to four tons of cargo or small vehicles within the fuselage, loading through a rear ramp and with room enough for two jeeps, a couple of Hawk surface-to-air missiles or a 105-mm howitzer, while a slung load of more than six tons can be carried on an under-fuselage hook, and all this with a maximum speed of almost 200 mph! As a ship-borne assault helicopter, the standard machines also have the invaluable feature of a boat-shape fully watertight hull.

A development of the CH–53A, the HH–53B, first flew in March 1967, as a combat rescue helicopter, equipped with a rescue winch, extra fuel tanks which can be jettisoned and an in-flight refuelling probe, with machine-guns, with an up-rated twin 3,080-shp General Electric T64–GE–3 turbo-shafts. Delivery of the first of eight such Super Jolly Green Giant helicopters for the United States Air Force's Aerospace Rescue and Recovery Service started in June 1967, and these were followed by almost sixty of the HH–53C version with

A Boeing-Vertol CH–47D Chinook carries fuel containers and an internal load as well, demonstrating its heavy-lifting capability.

engines up-rated to an even higher 3,435 shp.

The helicopter also continued to be developed further for its original troop-carrying and assault role, with the CH–53D for the United States Marine Corps being able to carry up to sixty-four fully-equipped marines, using twin 3,925-shp power-plants.

Other versions of the S–65 include the RH–53D mine-counter-measures helicopter for the United States Navy, of which thirty were built from 1972 onwards, while more recently triple-engined versions have been built for the United States Navy and the United States Marine Corps as the Super Sea Stallion. One of the largest customers for the S–65 series, however, is the Federal German Army, which operates more than 150 CH–53Gs built in West Germany by VFW-Fokker, although other users include the Israeli Defence Force and the Austrian Air Force.

A competitor to the S–65 on export markets is the Boeing-Vertol CH–47 Chinook, which was the winner of a 1959 design competition for a so-called "battlefield mobility" helicopter. Originally designated the Vertol 114, the helicopter was developed initially as the YHC–1B for service trials, intended to be a larger helicopter than Boeing-Vertol's 107, subsequently developed as the CH–46 Sea Knight for naval and marine use. After a first flight in September 1961, deliveries of the YHC–1B to the United States Army started the following summer, by which time all US designations had changed and the machine acquired the pre-production designation of YCH–47A. The pre-production and some early production versions of this large twin-rotor helicopter used twin tail-mounted 2,200-shp Lycoming T55–L–5 turbo-shafts, but most production machines have since used the more powerful 2,650-shp Lycoming T55–L–7 turbo-shafts. In common with the S–65 series, the CH–47s provide a rear-loading ramp for small vehicles, cargo or troops, carrying up to forty-four fully-equipped troops, depending on the version. In common with the S–65, the shp rating of the power-plants fitted has increased during the helicopter's production life from the original 2,200 shp through to 3,750 shp, and may of course rise still further. The United States Army standardized its medium and heavy-mobility helicopter fleet on the Chinook, operating a total of more than seven hundred of these machines, including many in South Vietnam. Later versions of the Chinook for the United States Army have included the CH–47B and CH–47C, with the latter being produced under licence by Agusta's subsidiary Meridonali in Italy for the Italian and Iranian armies, while the Royal Australian Air Force, the Spanish Army and the former South Vietnamese Air Force all received Boeing-built Chinooks.

Although the standard machine carries some three tons over a range of 240 miles at a speed of up to 190 mph, higher payloads can of course be carried over shorter distances, and range can also be increased at some cost to the payload. A heavy lift version of the twin-rotor concept, the XCH–62, is under development again by Boeing-Vertol.

However, vital though rapid transport across a battlefield must be, the American experience in Vietnam showed the need for a different type of helicopter from the small Iroquois and the large Chinooks, providing rapid fire support for the relatively lightly armed transport and assault machines as they attempted to land troops and supplies, or evacuate troops or casualties, from positions surrounded by heavy enemy troop concentrations. From this was borne the concept of the armed support helicopter.

Bell had originally investigated this concept as early as 1963, with the OH–13X Sioux Scout, a tandem-seat development of the Bell 47, which was flown in September, using a 260-hp Lycoming engine. When the United States Army decided on a crash programme to find and develop a helicopter urgently needed for the Vietnamese War, it seemed a sound idea to apply the same approach to an existing machine, but in this case the basis of the support helicopter was the UH–1 Iroquois, a proposal accepted by the United States Army in 1965. First flight of a prototype Bell 209 HueyCobra followed very quickly, in September 1956, with the same power-plant transmission system and rotor as the UH–1C version of the Iroquois, doubtless helping to speed development and lower development costs. However, much as the new helicopter owed to the Iroquois, there was little about its appearance which would suggest the connection to the layman, since the new helicopter used a very narrow fuselage with a tandem cockpit for a pilot and observer, reducing weight and leaving more payload available for armament, reducing drag and improving speed, and, most important of all since the frontal area was a maximum of just over three feet, reducing the helicopter's radar blip, making it a more difficult target to identify and aim guns or missiles at. Designated the AH–1G by the United States Army, that is, attack helicopter–1, 110 HueyCobras were ordered in spring 1966, with follow on orders bringing the total up to a staggering 1,100 machines by the end of 1972! Deliveries of the first production machines were started in June 1967.

The standard AH–1G HueyCobra uses an 1,100-shp Lycoming T53–L–13 turbo-shaft for a maximum cruising speed of just under 200 mph, and an operational radius of 230 miles, and its armament, effectively twice that of the standard Iroquois, includes either two

Miniguns or two 40-mm grenade launchers, or sometimes one of each, while stub wings provide armament strong points for XM–159 unguided rocket pods, or up to four XM–157 rocket pods, or an additional two Miniguns in pods comprising gun and ammunition. However, some variants have a turret fitted in the fuselage to fire a single M–61 Vulcan multi-barrel 20-mm gun, an XM–197 20-mm three-barrel gun or a 30-mm triple-barrelled gun, effectively modern and devastating reincarnations of the Gatling gun!

An up-rated twin-engined development of the HueyCobra for the United States Marine Corps has been the AH–1J Sea Cobra, using a 1,250-shp United Aircraft T400–CP–400 Turbo-Twin Pac engine. Deliveries of these helicopters started in 1970, but generally the armament and appearance is similar to that of the US Army versions, including armoured cockpit sides.

Taking the concept of the attack helicopter still further, the United States Army held a competition for an AAFSS, or advanced aerial fire support system, in 1965, with Sikorsky developing a one-off prototype, the S–66, specifically for this competition, and from this a further machine, the S–67, was developed for a 1972 United States Army competition for an Advanced Attack Helicopter, or AAH. Competition for the AAH was fierce, with several manufacturers producing excellent designs apart from Sikorsky's S–67, which for a period rejoiced in the name of Black Hawk. Bell produced the Model 309 or KingCobra as a further development of its work on the HueyCobra and SeaCobra, while Hughes produced its largest design to date, and Lockheed, almost entirely a manufacturer of fixed-wing aircraft, produced an interesting machine, the Cheyenne, and these aircraft, with the Sikorsky S–67, were prepared for a fly-off competition.

The Sikorsky S–67 flew for the first time in August 1970, utilizing a number of components from the S–61 series, but with a five-bladed rotor for improved performance and a reverse tricycle undercarriage enabling the helicopter to land tail-first, rather in the manner of the S–58 series. Establishing a world speed record for helicopters on 19th December 1970, of 220.9 mph, the S–67 could also handle an impressive armament, with up to sixteen TOW or HOT anti-tank missiles fitting under the stub wing strongpoints, while the armament also included such options as 7.62-mm Miniguns, cannon and grenade launchers, including provision for a chin turret, but even with all this, manoeuvrability aided by airbrakes on the stub wings, remained excellent. Two 1,500-shp General Electric T58–GE–5 turbo-shafts provided the essential propulsive power for this high performer.

In the search for a faster helicopter, the Lockheed XH–51A was a joint
Lockheed-US Army compound helicopter project.

Bell's Model 309 KingCobra consisted of a stretched HueyCobra,
with a larger diameter rotor blade. Two prototypes were built, with a
first flight in September 1971, and with existing HueyCobra and
SeaCobra customers in mind, the 309 used an up-rated version of the
United Aircraft Twin-Pac used on the SeaCobra, in the case of the first
prototype the 1,800-shp UACL T400–CP–400 Turbo-Twin Pac was
used for a maximum speed of up to 230 mph, although on the second
prototype, a more powerful Lycoming power-plant was used instead.
The 309's armament was intended to be somewhat lower than that of
the S–67, at eight underwing-mounted TOW or HOT missiles, plus 7-
62-mm Miniguns.

Lockheed had neglected the helicopter until the late 1950s, possibly
being attracted to the concept at this time after the company's position
in the civilian airliner market had waned following the success of the
Constellation series and the relatively poor commercial achievement of
the Electra. However, while the growing interest in the helicopter and
the advances in helicopter capabilities also made the venture into
rotary-wing aircraft seem more worth while, Lockheed introduced
some significant technical advances of its own after turning to the
helicopter in 1958. Lockheed's interest in helicopters took two forms,
firstly being interested in the potential for higher speeds presented by
the compound helicopter, and secondly paying attention to the
principle of the rigid rotor, for improved manoeuvrability. After

experiments with the CL–475 prototype in 1959 and with a larger development of this machine, designated the CL–595, of which three were built for trials by the National Aeronautics and Space Administration and both the United States Navy and the United States Army with the designation XH–51A, the manufacturer moved towards its large AH-56 Cheyenne attack helicopter. The Cheyenne incorporated both a rigid rotor and a pusher propeller at the tail to provide additional forward propulsion, while also replacing the conventional tail-mounted rotor. Using a single General Electric T64–GE–16 turbo-shaft of 3,925 shp, this machine could attain a maximum speed of 244 mph, but even this high speed was lower than that achieved by the turbo-jet-assisted CL–595, which managed 272 mph on one occasion!

The Cheyenne prototype was ready for the US Army's AAFSS competition, in which it was pitted against the Sikorsky S–67, and won, with the US Army ordering ten development machines, although only four were delivered before the contract was suspended. Curiouser still, the United States Army ordered 375 Cheyennes in January 1968, and then cancelled this order a little over a year later.

The fourth American attack helicopter at this time was the Hughes AH–64, a twin 1,536-shp General Electric T700–GE–700 turbo-shaft-powered helicopter capable of carrying up to sixteen anti-tank missiles on its under-wing strongpoints as well as Hughes XM–230 30-mm chain gun mounted on an under-fuselage 'chin' turret, and capable of a maximum speed of just under 200 mph and a range of 360 miles without auxiliary fuel tanks.

The Lockheed AH–56A Cheyenne was an unsuccessful contender for a US Army attack helicopter contract.

When the Advanced Attack Helicopter Competition was held in 1972, the Lockheed Cheyenne, the only compound helicopter in the competition and with the benefit, one would have thought, of having 90 per cent of all power diverted to the tail rotor in flight, was eliminated, and eventually, after further consideration, the Hughes AH–64 was selected instead of the Bell 309, which was designated the YAH–63 by the United States Army, in 1976, some four years after the competition started. The helicopter underwent further development flying with pre-production YAH–64s evaluating systems for use in the production helicopters, which started to roll off the assembly line in 1980, with some six hundred machines likely to be built during the early 1980s.

Sikorsky, unfortunately, suffered a further mishap when the S–67 Black Hawk prototype crashed during a particularly spectacular demonstration at the 1974 Farnborough International Air Show in Britain, killing one crew member and fatally injuring the other. The name Black Hawk, however, was retained and re-used on a new

The Hughes AH–64 succeeded in winning the US Army order.

Three Hughes YAH–64 prototypes in formation.

helicopter for the United States Army, developed as the Sikorsky S–70, and intended to replace the large number of Bell UH–1 Iroquois in Army service.

Meanwhile, the Soviet Union was not allowing itself to fall behind, indeed, as so often happens with military technology, as opposed to civilian, the Soviet Union was at least abreast of developments in the United States, if not ahead, and of course nowhere could this Soviet trait be more clearly shown than in helicopter development, which has become the Soviet aircraft designers' forte.

The first appearance to Western observers of the main Soviet attack helicopter, the Mil Mi–24, known to NATO as the 'Hind', was in 1972, and at first this was thought to utilize many components from the Mil Mi–8 transport and assault helicopter, but subsequent observation has indicated that the Mi–24 is in fact a largely new helicopter. Although superficially conforming to the basic western, and that really means American, concept of the attack helicopter, the Mi–24 is slightly larger, and behind the standard attack tandem two-seat cockpit is a narrow cabin for about eight fully equipped combat troops. The single-rotor Mi–24 is powered by a twin 1,500-shp Isotov turbo-shaft installation, giving a maximum speed of around 180 mph, although stripped-down variants are believed to be the A–10 helicopters in which the Soviet Union has achieved speeds of up to 210 mph. There are six strongpoints on the stub wings for rockets or missiles, while the latest version known as 'Hind D', has a modified front fuselage with a chin turret for a 20-mm or 30-mm Gatling-type four-barrel machine-gun. The original 'Hind A' is not thought to have entered service in any substantial numbers, while 'Hind C' is relatively lightly armed, operating possibly in the same way as an Iroquois with rocket pods might, and 'Hind D' is the standard helicopter for anti-tank operations and heavy counter-insurgency work. Surprisingly, few, if any, Mi-24s have entered service with the other Warsaw Pact nations, although a few have been supplied to Libya, and a substantial number are based in East Germany with the Soviet Frontal Aviation forces there.

To a great extent, until recently the attack helicopter has been largely confined to the American and Russian aircraft industries, with the French abandoning plans to develop an attack version of the Anglo-French Westland-Aerospatiale Lynx helicopter, and only recently has this tendency been broken with the advent of the Agusta A.129 from Italy.

So far, the battlefield helicopter as such has played a relatively limited role in warfare, thankfully because major set piece battles are far rarer than counter-insurgency operations and the skirmishes between relatively small groups of men. The helicopter, for various reasons, has also played a relatively small part in many of the conventional wars which have taken place over the period since the end of World War II. It has, for example, been little used in the several wars between the Arabs and the Israelis, with its role largely confined to combat rescue operations and liaison.

However, helicopters were used to a limited extent, simply because they were also in limited supply, when Turkey invaded Cyprus in 1973, after a *coup d'état* on the island prompted fears over the security of the Turkish Cypriot community. A Turkish fleet sailed for Cyprus on 19th July, with an estimated 70,000 troops and forty tanks aboard a fleet of landing-craft, and although the Greek fleet was ordered to sea at this time, distance and the proximity of Cyprus to Turkey restricted their actions. On landing in Cyprus, the small number of Bell UH–1 Iroquois helicopters in service with the Turkish Army was used on communications, liaison, observation and CASEVAC duties, and for some limited troop movement.

Probably a more notable application of the helicopter came during the same Cyprus emergency, but in a completely different way. After a cease-fire was arranged on 22nd July, the opportunity was taken to evacuate some 1,500 British residents and holiday-makers living in the Turkish-occupied parts of Cyprus, mainly around Kyrenia on the north of the island. The British commando carrier, HMS *Hermes* a sister ship of HMS *Albion* and *Bulwark*, was ordered to the area, and on 23rd July, the ship's two Westland Sea King anti-submarine helicopters and the eight Westland Wessex commando helicopters also aboard, picked all 1,500 civilians off the beaches and flew them aboard the ship, whence they were taken to the Royal Air Force sovereign bases on the other side of the island and flown back to Britain on RAF transport aircraft.

A more substantial example of the helicopter in conventional warfare came in 1979, after a Christmas invasion of Afghanistan by Soviet forces, who installed a puppet regime which immediately met with fierce resistance from relatively lightly armed and disorganized Afghan tribesmen. Supported by Mil Mi–8, Mi–10 and Mi–24 helicopters, the Soviet forces quickly neutralized the Afghan Army and Air Force, with the helicopter contributing greatly to the speed of the Soviet take-over.

Post-invasion resistance to the Soviet occupation by Afghan tribesmen, mainly gathered in the mountainous south-east of the country close to neighbouring Pakistan, has resulted in ground-attack operations by Mi-24 helicopters, using rockets and machine-guns and, on occasion, chemical weapons, including grenades, in spite of the resistance forces being armed only with obsolete infantry weapons. The country was soon dotted with beacons to aid helicopter operations at night, and according to some recent reports, few of these beacons seemed to have suffered serious attack by the tribesmen, suggesting either a failure to appreciate their value or a gradual weakening of their ability to launch even minor offensives. While the Afghan situation is in many ways more akin to a counter-insurgency operation, it was preceded by a helicopter-supported invasion, and it is the invader who is seeking to establish a permanent presence and the oppression of resistance, rather than the national government, with its own armed forces in considerable disarray.

The first real application of the helicopter in an all-out war between two nations came with the outbreak of hostilities between Iran and Iraq on the morning of Wednesday 9th April 1980. Three Iraqi helicopters, of unknown type, made a machine-gun attack on Iranian positions in the vicinity of Qasr Shiria in the central border area later during the first day, at about 14.00, following fierce fighting between ground forces. The pace of the conflict quickened, and by 24th April, Iranian helicopter gunships with top cover provided by McDonnell Douglas F–4 Phantom jet fighters, were attacking Kurdish guerillas operating in support of Iraqi troops, mainly in and around the Iraqi city of Sanandaj, with early reports from the scene of the battle suggesting that some fifty civilians were killed by the helicopter attack on the city. However, after the war had developed, a stalemate was eventually reached, with relatively little offensive activity by either side after late 1980 and throughout 1981 or most of 1982.

One reason for the stalemate was the uncertain attitude of the Soviet Union towards the new Iranian revolutionary government. It wished to pull Iran into the Soviet sphere of influence and was therefore reluctant to support Iraq, which was at the same time opposed to the main Soviet ally in the Middle East, Syria. Iran, meanwhile, had suffered a considerable reduction in the effectiveness of its armed forces after the overthrow of the pro-American rule of the Shah, and the installation of an anti-American revolutionary government which had either connived at the seizure of the United States Embassy in Tehran, and the holding hostage of the American diplomats, or at least was unwilling to exert any authority over the revolutionary 'students' who were holding the diplomats in breach of standard diplomatic convention.

The seizure of the fifty-three American hostages was in itself a crisis,

167

but at first the Americans put their faith in diplomacy as an answer to the problem. However, after 4th November 1979, the situation continued to deteriorate, and eventually stronger measures than diplomacy or an arms embargo had to be considered.

The world's largest aircraft-carrier, the nuclear-powered (CVN), USS *Nimitz* arrived outside the Persian Gulf on 25th April 1980, with a complement of aircraft and men which included eight large Sikorsky S–65 helicopters and ninety commandos drawn from all four branches of the American armed forces. The mission was fated from the beginning. After take-off on the still secret mission, the rescue of the hostages, one of the S–65s developed engine trouble and was forced to return to the aircraft-carrier, while a second made a forced landing in the Iranian desert, with its crew being picked up by another helicopter. On landing at a remote desert rendezvous some 260 miles south-east of Tehran to refuel from six United States Air Force C–130 Hercules transports, the assembled group of transports and helicopters were seen by a busload of some fifty Iranian civilians, forcing the Americans to hold the Iranians hostage to retain the secrecy of their mission. A third helicopter developed trouble at this stage, taking the force below the essential minimum of six helicopters regarded as essential for the success of their mission, and forcing the commander of the small force to call off the rescue attempt. At this stage, it would seem that some confusion arose, with a C–130 taking-off, but colliding with a helicopter preparing to take-off and hidden by the dust raised by the whirling rotor blades, killing eight American servicemen and injuring four others. The decision was then taken to abandon the five remaining helicopters, and the force withdrew in the remaining five Hercules transports.

The price of this failure was a major propaganda triumph for the Iranians, with the burnt corpses of the Americans killed in the accident being paraded before newspaper and television cameras, while the hostages themselves were scattered and not released until early 1981.

Inevitably, there were an inquiry about the failure of the mission, accompanied by reports about the lack of special maintenance for the helicopters preparing for the mission, and even one allegation that the helicopters were accidentally drenched by fire hoses before the mission departed.

So much for the short history of the helicopter on the battlefield. However, the main area for future use of the battlefield helicopter must be the so-called Central Front of the North Atlantic Alliance, effectively the entire frontier area of West Germany from the Baltic to the Swiss border, and the territory for some miles on either side, extending into the Netherlands and Belgium as well as northern France, and East Germany and Czechoslovakia on the eastern side of the border. It is barely imaginable that NATO would fight an offensive campaign, or move far towards the east even after containing and turning a Soviet invasion, so this effectively limits the battle area to Western Europe.

Some idea of Soviet tactics with battlefield helicopters has reached the West. In an article in *Aviatsiya i Kosmonavtika*, reported in the British magazine, *Flight International*, an assault supported by Mil Mi–8s is described. In an exercise to land troops some 8,000 feet up in the mountains, in an operation reminiscent of some of the early helicopter-borne troop movements of the Korean War, the assault is preceded by an artillery barrage or fighter-bomber strikes, perhaps both, with the helicopters readied for the assault three minutes before the barrage is lifted. The helicopters move in to land their troops, flying to the landing zone in pairs, with the rest of the helicopters providing supporting fire to suppress any moves by the defenders. As helicopters disembark their troops, they join the other helicopters in providing supporting fire for the helicopters landing troops.

Mil Mi–24 'Hind' tactics are also described, with these helicopters operating very much like the old concept of the dive bomber or fighter-bomber attacking enemy strongpoints and ammunition dumps with rockets, bombs or grenades, and machine-gun fire.

A *Red Star* article, also reported in *Flight International*, describes the way in which a pair of Mi–24s would break up an enemy tank formation. Effectively, the Mi–24 tactics include climbing steeply before launching anti-tank guided missiles, while elsewhere in the operation, helicopters bomb enemy units whilst flying at high level, and fire unguided rockets.

A further interesting point from these articles is that the Soviet helicopter units can use portable cranes to remove engines from damaged helicopters, enabling an engine change to be made without removing the main rotor and, even more important, without support from self-propelled heavy cranes, which may be unable to reach the downed helicopter over difficult terrain.

The need of the Mi–24 to climb before launching its rockets or missiles does not necessarily indicate a desire for altitude, but more probably simply means that Soviet helicopter crew, in common with their western counterparts, use the terrain to 'hide' their helicopters, waiting until the last possible moment to rise from behind the smallest cover provided by the terrain, or perhaps a small wood or copse, and then to attack while the element of surprise is theirs. This obviously requires training, but in practical terms it also dictates that helicopter crews spend some considerable time working over the area in which they are likely to fight, looking for every small depression, a sunken road or even a river bank, let alone anything larger, behind which they might conceivably hide their helicopter. This does give the defender some element of advantage over the aggressor, but in military terms a three-to-one ratio in favour of the attacker is regarded as the barest minimum if success is to be possible, and so the defender naturally has to face overwhelming enemy strength.

Such developments in tactics also force developments in equipment, and modern helicopters are being fitted with periscopic sights with which to see over the undulations in the terrain. The problem of the rotor interfering with any periscope is overcome by placing the head of the periscope above the rotor hub using modern optical and electronic technology to transmit the picture to the observer in the helicopter. Even with such cover and the aid of such brilliant equipment, and even allowing for the considerable noise of an advancing tank formation, the helicopter still suffers from considerable rotor noise, risking betrayal of its position to an enemy. Current development is thus being concentrated on quieter helicopters.

A Hughes OH–6A with the new periscope sights for improved vision while the helicopter is using the available cover concealment.

Some Soviet Mil Mi–24 helicopters are allocated the task of helicopter-to-helicopter combat, realizing that opposing tank armies are likely to be accompanied by opposing anti-tank helicopter formations. This does not imply a stalemate, since differing strengths, the element of surprise, morale, and the quality of men, training and equipment all play a part, but it does present a problem, and a lack of advantage which has to be broken. The German Luftwaffe plans to use its Alphajet advanced jet trainers in the anti-helicopter role in any major conflict, but at present none of the other major military powers seems to be adopting this tactic. Indeed, the British plan to use some of their trainers in the air-to-air defence role, where they may be hopelessly outclassed, rather than in the anti-helicopter role where they could be used to devastating effect. Few countries have adopted attack helicopters as such, but again, armed jet trainers can be effective in the anti-tank role, and aircraft such as the Harrier can also be extremely effective in a variety of tactical roles. However, most nations likely to fight a major modern land battle do equip their helicopters with anti-tank missiles, even designating helicopter units to this role, although by nature of their design and commonality with the helicopters, usually of the same type, in the assault and support roles, they can be switched to other roles quickly. For all but a few air forces and air arms, flexibility is more important than possessing specialized attack helicopters, and as we have seen, even the Soviet Mi–24s can be used in the assault and CASEVAC roles, amongst others.

All of this suggests that a future war, even if fought with conventional weapons, would be bloodier and more furious than even World War I. It remains to be seen whether the helicopter has at last made the tank obsolete, but certainly no major army takes this view, with every army of any note retaining armoured units as either the spearhead of an assault, or as a major element in breaking any assault.

8

ENTER THE SMALL-SHIPS HELICOPTER

Obviously, the helicopter was perfectly able to operate from aircraft-carriers of any size, right from the start, with the only possible constraint being the question as to whether or not the somewhat ungainly new type of aircraft could fit neatly onto a carrier's lifts; even this problem was soon solved with collapsible rotors and folding tailplanes. However, even during the early experiments, the pioneers had it in mind that the helicopter should be able to operate from small warships and merchant vessels, bringing not only "over-the-horizon" surveillance to all but the smallest vessels, but also giving the fleet or convoy instant and highly manoeuvrable anti-submarine protection. It was realized that the helicopter could free the convoy from many of the perils which awaited it once outside the range of shore-based maritime-reconnaissance aircraft, and without the need to allocate scarce and costly aircraft-carriers with their air groups to often quite small convoys. Essentially, the concept offered greater flexibility, with convoys easier to form and smaller but more frequent convoys becoming possible. Paradoxically, the smaller the convoy, the more difficult it would be for the enemy to find, and the less attractive a target it would make once located, while the smaller number of ships also reduced the possibility that the convoy would include older and slower vessels, so convoys could also become faster. For the enemy submarine, it presented a vicious circle, but for the naval commander under pressure to keep shipping on the move and in safety, the opportunity presented by the helicopter seemed to be almost miraculous.

This said, it seems that there were long delays before the first suitable small warships could be found, and there was also some delay before helicopters suitable for small-ship operation appeared. The helicopter really required a more stable platform than the narrow but fast World War II escort vessel could provide, although that said, the high speed of the warship became less important due to the speed of the helicopter, and it is worth noting that modern NATO escort vessels, the destroyers, frigates and corvettes, are in fact considerably slower and broader-beamed than their World War II counterparts. The need for some form of hangarage in addition to a landing platform also meant that a fair amount of space was required, and many navies were reluctant to see one or two turrets disappear from their destroyers, but in the end, the more modern vessels have a lower armament and are larger than their World War II counterparts. To be fair, however, the escalation in warship size has also been due to the growing sophistication of the radar and missile-control systems, and to the need to provide professional, as opposed to conscript, crews with a better standard of accommodation, the helicopter should not be seen as the sole culprit.

It is important to understand that the early helicopter was not always suitable for small-ship operation. Helicopters, such as the successful Sikorsky S–55 and its British counterpart, the Westland Whirlwind, were not particularly stable on a heaving deck of small ship, and they were also too big for the smallest vessels. The Sikorsky S–58 and the Westland Wessex provided the stability which had been lacking, but their size still meant that they remained as options only for the larger warships, such as the British County class of guided missile destroyers. It took the right sort of helicopter to stimulate warship designers and their naval overlords, and the right sort of warship to make helicopter manufacturers feel that they stood a chance of gaining a worthwhile naval order. Truly a chicken and egg situation.

Two of the early efforts to cut the helicopter down to escort vessel size came from Agusta in Italy and from Westland in Britain, with respectively the Agusta A.106 and the Westland Wasp.

The A.106 was the successor to a number of small piston-engined lightweight helicopter prototypes built by the Italian manufacturer, and which ultimately led to the turbo-shaft-powered A.105, which made its first flight in April 1964. A prototype A.106 followed in November 1965, and this machine was ultimately put into production in fairly limited numbers, just twenty production machines were built, to operate as an anti-submarine 'killer' helicopter from the Impavido-class destroyers. A single 330-shp Turbomeca Astazou turbo-shaft provided the small twin-seat helicopter with a maximum speed of about 100 mph, and the ability to carry up to two Mk. 44 homing torpedoes. A Bell twin-bladed rotor indicated the continuing connection between the American and Italian companies, even though Bell has not at any time produced such a small specialized helicopter itself. Unusually, in the light of the approach taken with the Westland Wasp, the A.106 used a twin-skid undercarriage.

The Westland Wasp proved to be a larger helicopter, being based mainly on the Scout helicopter for the British Army, mentioned in the previous chapter. Wasp development took place at the same time as

173

that of the Scout, with one of the two Saunders-Roe P.531 prototypes taking part in extensive naval trials during late 1959 and 1960, and being joined by two further prototypes equipped with a long-stroke castoring undercarriage in place of the skids on the Army machines. The long stroke undercarriage was adopted for the production machines, offering not only greater security against heavy landings on a pitching warship, but easier deck handling. The first flight of a production Wasp was on 28th October 1962. Apart from the undercarriage, the Wasp differed from the Scout in having a slightly more powerful, 710-shp as against 685-shp, Rolls-Royce Nimbus turbo-shaft, and certain tailplane modifications, but of course the weapon load also differed, with two Mk. 44 homing torpedoes replacing the AS.11 or AS.12 missiles more usually found on the Army machines.

Most significantly, the Wasp proved to be the first really big success for Westland Aircraft, with more than eighty being delivered to the Royal Navy for use on the shipboard flights of the Leander, Tribal, Rothesay and Amazon-class frigates, and others being supplied to the Royal New Zealand Navy, the Indian Navy, South African Navy, Chilean Navy and the Royal Netherlands Navy, and helping to establish the manufacturer with a growing reputation in the design and development of small naval helicopters.

One of the first small-ships helicopters was the Westland Wasp, which could carry two anti-submarine torpedoes under its fuselage; this is a Royal Netherlands Navy machine.

A Royal Navy example of the Wasp, with torpedoes.

Even a Wasp is a tight fit aboard a frigate.

The Tribal-class frigates of the Royal Navy were amongst the first small warships anywhere to be designed and built with the helicopter in mind. Indeed, the first of the class, HMS *Ashanti*, 1,700 tons, held the dual distinction of being the first warship to be designed and built to operate an anti-submarine helicopter and the first to operate with COSAG, or combined steam- and gas-turbine propulsion. Completed in 1961, *Ashanti* was followed by six sister ships, the last of which, HMS *Tartar*, was commissioned in 1964. Unusually for Royal Navy vessels, the Tribal class were single-screw, but were able to operate as general-purpose frigates, replacing destroyers and with a considerable emphasis on self-sufficiency for distant water 'policing' operations, for which, of course, a helicopter would be useful. Unlike later Royal Navy helicopter-carrying warships, the Tribal-class featured a collapsible hangar rather than a more substantial fixed structure.

Conversion of most of the Whitby class and all of the generally similar Rothesay-class frigates followed, with a reduction in their anti-submarine armament in order to accommodate the new and rather more flexible helicopter. Conventional twin-screw steam turbine vessels built during the late 1950s and into the early 1960s, these two classes were renowned for their excellent sea-keeping qualities and presented an appearance of purposefulness, while the classes were ordered by New Zealand, South Africa and Australia, as the latter's River class, albeit without helicopters.

Both the Rothesay or Whitby classes and the Tribal class also boasted Short Seacat surface-to-air missiles in addition to their 4.5-inch and light anti-aircraft gun armament. This pattern was followed in the next class of frigates, a direct development of the Rothesays, the famous Leander class, which was designed to operate helicopters as both general-purpose and anti-submarine frigates, with no less than twenty-six ships of this class built for the Royal Navy itself, with sixteen of the original design and ten being the larger broad-beamed version, commissioning between 1963, with HMS *Leander* and 1973, with HMS *Ariadne*. The Leander-class ships were more modern and larger frigates than the Rothesays, and the broad-beamed version could take a second quadruple Seacat launcher. British-built Leanders were supplied to Chile and New Zealand, while the two last ships of the River class followed the Leander hull form, but were built in Australia. The class was also built in India, and the Royal Netherlands Navy's Van Speijk class was also based directly on the Leander design.

However, before the Leanders entered service, a much larger British helicopter-carrying destroyer class was introduced, the County class of eight 6,200-ton full-load displacement guided missile destroyers. The County-class ships also used COSAG propulsion, but with twin screws unlike the Tribals, and were armed with short-range Short Seacat and longer-range HSD Sea Slug surface-to-air guided missiles, both of which retained some secondary surface-to-surface capability. The County class were designed with a hangar and landing platform for a single Westland Wessex anti-submarine helicopter. Originally intended to be light cruisers, the official destroyer designation was given to the class after the decision was taken to complete these ships with 4.5-inch destroyer main gun armament rather than the 6-inch guns of a light cruiser.

The big advance of this class was the ability to operate a fairly large helicopter with its own dipping sonar, since all smaller shipboard helicopters notably the A.106 and the Wasp, could only act as killers, being dependent upon visual sighting of a submarine or on the sonar of the mother ship. However, the addition of the Wessex was made at some cost in over-design of the County class, with a hangar tailored to accommodate a helicopter not just of Wessex size, but also of Wessex shape, with access to the hangar being from the side. This meant that in years to come, smaller helicopters had to replace the Wessex, with a subsequent downgrading of capability, rather than the ship taking the logical Wessex successor, the Westland Sea King.

The first of this elegant class, HMS *Hampshire*, was commissioned in 1962, and was followed by her seven sisters, with the last being commissioned in 1970.

A successor to the County class, the 'one ship class' HMS *Bristol*, originally designed to be a class of several ships to escort a new class of attack carrier, but subsequently completed as a trials ship for new weapons systems after the carriers were cancelled during the mid-1960s, also used COSAG propulsion and had a landing pad for a single Lynx helicopter, but no hangar.

Another significant British helicopter-carrying warship class of this period was the Tiger class. These three cruisers had been laid down late in World War II, and work was suspended on them after launching. During the late 1950s, they were re-designed and completed to enter service during the early 1960s, and renamed the Tiger-class. It was eventually decided to convert these vessels to carry helicopters, although before this programme could be completed, HMS *Lion* was withdrawn from service and subsequently scrapped in 1972. Nevertheless, her two sisters, HMS *Tiger* and HMS *Blake* (the Royal Navy seems to have run short of big cat names, possibly due to the existence of the Leopard-class frigates at this period) were converted

177

with a hangar and flight-deck for up to four Sea King helicopters in place of the after twin 6-inch gun turret and twin 4.5-inch gun turret. The class thus offered a considerable anti-submarine capability, with the ability to act as flagship for a task-force.

The growing number of warships able to handle helicopters also affected the Royal Navy's training procedures, with a Royal Fleet Auxiliary, *Engadine*, built and commissioned in 1967 as a helicopter support ship, with one of her roles being the training not only of Fleet Air Arm helicopter pilots, but also of flight-deck personnel and air-traffic controllers. *Engadine*, a diesel vessel of 9,000 tons and with a cruising speed of just 14.5 knots, could accommodate up to four Wessex helicopters or two Sea Kings and one or two Wasps, with a fairly large landing pad and her own hangarage. The ship has also been used for amphibious exercises.

Other vessels of the Royal Fleet Auxiliary, the merchant navy-manned fleet train for the Royal Navy, were also being built with the capability to handle helicopters. Obviously, the concept of self-contained anti-submarine protection for these vessels and any others within a convoy was much in mind in providing for helicopters, but there was another role as well, that of 'vertical replenishment', using the helicopters in effect as flying cranes to transfer stores from an RFA to a warship under way. Amongst such vessels were the "O1" class of large fleet tankers, with *Olwen* and her two sisters built in 1965, 1966 and 1967, and each of these 36,000-ton displacement vessels able to accommodate up to four Sea King helicopters. Another class of tankers, the Tide class could also handle up to six helicopters, but in 1979 a new class of two ships without helicopter facilities, the first of which was the *Appleleaf*, was purchased for the Royal Navy, largely because these vessels were bought while under construction for a merchant shipping line, which no longer required them while the Royal Navy needed new large tankers urgently. The small fleet tankers of the Rover class, built between 1969 and 1974 can each accommodate one Sea King, while the large fleet replenishment vessels, *Fort Grange* and *Fort Austin*, 23,600 tons, completed in 1978 and 1979 respectively can each take up to four Sea Kings, and the smaller stores support ships, *Stromness* and *Tarbatness*, 14,000 tons and commissioned in 1967, can handle a single Sea King. Two other ships, RFA *Resource* and *Regent*, 22,890 tons, take two Sea Kings.

The Sea King was the Westland licence-built development of the Sikorsky S–61. Originally designed to meet a United States Navy requirement for an amphibious helicopter capable of filling both the hunter and killer roles, Sikorsky was awarded a contract to build

A USN Sikorsky Sea King on a humanitarian mission.

HMAS *Melbourne*, Australia's much modernized aircraft-carrier, with four Westland Sea Kings with her attack squadron of McDonnell Douglas A–4 Skyhawks.

A flight of Mitsubishi-built
Sikorsky HSS helicopters.

A Royal Navy Sea King
retrieves its dunking sonar.

One of the Federal German Navy's search and rescue Westland Sea Kings.

The Sikorsky S–65 is used in a variety of roles, including rescue, assault and mine counter-measures.

prototypes in September 1957, with a first flight following in March 1959, as the XHSS–2, followed by seven pre-production YHSS–2s for service trials in 1960. Service deliveries of the production HSS–2 to the United States Navy commenced in 1961, and the designation was changed the following year to SH–3A Sea King, as part of the overall standardization of US service aircraft designations in 1962.

Early production versions of the Sea King used twin General Electric T58–GE–8B turbo-shafts of 1,250-shp each for a maximum speed of 150 mph and a range of up to 600 miles; these machines were all equipped with dunking sonar and able to carry up to four homing torpedoes or depth-charges for anti-submarine duties. A number of SH–3As were delivered to Canada, with thirty-seven out of a total order for forty-one CH–124s, to use the Canadian designation, being assembled in Canada, while Mitsubishi in Japan contracted to build more than eighty for the Japanese Maritime Self-Defence Force. During the mid-1960s, the SH–3D was introduced as the standard production version, using more powerful 1,400-shp General Electric T58 turbo-shafts, and this version has been built under licence by Agusta in Italy and by Westland Aircraft in the United Kingdom, although the Westland version uses more powerful Rolls-Royce Gnome H–1400 turbo-shafts of 1,500 shp each, and more advanced British avionics than the American and Italian aircraft. Westland and Agusta versions share the Sikorsky Sea King name. The Westland versions have been built in quantity for the Royal Navy and, without the dunking sonar, for the Royal Air Force's search and rescue flights, as well as for the Royal Norwegian Air Force, the Federal German Navy, for the Indian and Pakistani Navies, the Belgian Air Force, the Royal Australian Navy and for Egypt. A transport version, the Westland Commando, is also now in service with the Royal Marines and with the Egyptian Army.

Other developments of the Sikorsky S–61 followed, including some highly successful commercial versions of what must be the most successful medium-sized helicopter to date. A number of the early SH–3As were converted by the United States Navy for mine counter-measures, MCM, duties, but ultimately it was decided that a larger helicopter based on the S–65 was needed for this duty, while others became utility machines with their ASW equipment removed and 7.62-mm Minigun pods fitted. However, other versions also included the famous 'Jolly Green Giant' CH–3C or HH–3Es of the United States Air Force. These machines differed from the original in having a re-designed rear fuselage with a ramp and a kneeling undercarriage, and while early versions used 1,300-shp T58 engines, eventually all of the

133 machines put into service by the USAF were fitted with 1,500-shp versions of the T58. Initially intended as medium transport helicopters, the S–61s in USAF service gained fame after being placed with the USAF's Aerospace Rescue and Recovery Service, fitted with rescue hoists, flight-refuelling probes with which to refuel from a KC–130 Hercules transport, and protective armour and designated HH–3E. On entering service with the USAF in Vietnam, wearing a distinctive green camouflage, these machines became known as the "Jolly Green Giants", a reference to the trade mark of a brand of sweet-corn which, apparently, figured prominently in US Army and Marines rations during the Vietnam War. An idea of the capabilities of the HH–3E can be gathered from the fact that two of these made the first non-stop helicopter crossing of the North Atlantic, leaving New York on 31st May 1967, and arriving at the 1967 Paris Air Show on 1st June, after flying for 4,270 miles, each aircraft refuelling in mid-air nine times during the crossing. During the Vietnam War, HH–3Es frequently refuelled from Lockheed KC–130E Hercules tanker aircraft. A version of these helicopters, designated the HH–3F, was delivered to the United States Coast Guard, which has about forty, but the Aerospace Rescue and Recovery Service discovered that a more powerful machine was needed, and has since standardized on the larger H–53, or S–65.

A USN CH–46 Sea Knight makes a delivery aboard an American aircraft-carrier.

Similar in appearance to the S–61 is the Sikorsky S–62, a smaller helicopter with an amphibious hull. An earlier design than the S–61, the early S–62 used many of the technical features of the S–58 but with a single 1,050-shp General Electric T58–GE–6 turbo-shaft de-rated to 670 shp on the prototype, which first flew on 22nd May 1958. Commercial deliveries commenced in July 1962, but one of the best-known customers for this helicopter must be the United States Coast Guard, which has received about a hundred S–62As, designated the HH–52A and using a 730-shp General Electric T58 engine as a dedicated search and rescue helicopter. A small number of S–62s have been exported to military and naval customers, but this attractive helicopter has never quite enjoyed the success of its larger cousin, the S–61, or the earlier S–55 and S–58.

A contemporary of the S–61 was the Boeing-Vertol CH–46A Sea Knight, another member of this family of twin tandem-rotor helicopters. The Sea Knight had its origins in the manufacturer's Vertol 107 project, first mooted in 1956, with two H–21 helicopters modified to take twin gas-turbine installations the following year, with the prototype Vertol 107 making its first flight on 22nd April 1958, using two 860-shp Lycoming T53 turbo-shafts. The manufacturer foresaw the main application for the helicopter coming from the civilian market, and the original concept was for a twenty-three passenger and three-crew helicopter, although the military and naval uses of the machine were certainly not overlooked. United States Army interest led to the ordering of ten YHC–1A pre-production aircraft, but this was cut back to three aircraft after the Army decided to pursue development of the larger CH–47 Chinook instead. Commercial development of the 107 series was not neglected during this period, however, but the next indication of service interest came in 1961, when a modified version known to the manufacturer as the 107M, was selected by the United States Marine Corps as their HRB–1 medium assault helicopter. In common with the S–61, the marine 107 soon enjoyed a change of designation, becoming the CH–46A Sea Knight in July 1962, and after an initial order of fourteen machines, the USMC soon ordered a further six hundred, many of which were the ultimate production version, the CH–46D.

Powered by twin General Electric T58–GE–10 turbo-shafts of 1,400-shp each, the CH–46D, and the similar CH–46F which includes additional avionics, offered a maximum speed of 157 mph and a range of more than seven hundred miles, with a crew of three and up to twenty-five fully equipped troops or fifteen stretchers and two attendants. In common with many transport helicopters of what may be described as the third generation, a rear-loading ramp was provided for vehicles and equipment and to assist with striking down into carrier hangar decks, there is power folding of the rotor blades. A small number of helicopters, including twenty-four UH–46As and some UH–46Ds, were ordered by the United States Navy, mainly for use aboard its Fast Combat Support Ships providing vertical replenishment of supplies between these and warships at sea.

Kawasaki in Japan obtained a licence to build the 107 series, and amongst those delivered to the Japanese Self-Defence forces were a number of KV–107/II–3s for mine counter-measures with the Maritime Self-Defence Force and thirty SAR machines for the Air Self-Defence Force. A number of the Japanese-built machines used Rolls-Royce H–1200 Gnome turbo-shafts, and a mixture of both Japanese and Boeing-Vertol-built Gnome-powered 107s were supplied to the Royal Swedish Navy as their HKP 4 anti-submarine helicopter, with twelve of these, equipped with dunking sonars, being supplied during the late 1960s and early 1970s to provide the backbone of the RSwN's anti-submarine coastal forces. Kawasaki has held, since 1965, the exclusive rights to manufacture and sell 107s to customers outside North America. The only other major user of the 107 series is the Canadian Armed Forces, with these machines operating as CH–113 Labrador SAR aircraft, and CH–113A Voyageur transports.

Yet another helicopter operational during this period with the United States Navy has been the Kaman H–2 Seasprite, which was the winner of a design competition in 1956 for a fast, long-range utility and rescue helicopter with all-weather capabilities. The first of four prototypes flew for the first time on 2nd July 1959, using a single 1,025-shp General Electric T58–GE–6 turbo-shaft. An initial production run of eighty-eight UH–2A machines started to enter service in mid-December 1962, using a 1,250-shp T58–GE–8 turbo-shaft to carry up to eleven passengers in addition to a crew of three. A second batch of 102 UH–2Bs followed using the same power-plant. The UH–2 Seasprite was the first Kaman helicopter to abandon the interconnecting rotors for a conventional rotor; this attractive helicopter also featured a retractable undercarriage. It could fly at speeds of up to 165 mph and operate over a normal range of 445 miles. Typical duties for the UH–2A and UH–2B included carrier plane-guard duties, liaison, communications, rescue, CASEVAC and some carrier or escort vessel on board delivery operations. In 1965, the feasibility of converting the Seasprites to twin-engined power was examined, and after successful trials, a programme was initiated in 1967 for the conversion of all UH–2A and UH–2B aircraft to UH–2C

186

The sleek lines of the Kaman Seasprite.

standard, using two 1,350-shp General Electric T58–GE–8F turbo-shafts. While US Army trials with a handful of Seasprites, known in military service as the Tomahawk, did not produce any worthwhile orders, the US Navy did nevertheless introduce a version known as the HH–2C with armour and fitted with a chin-mounted Minigun turret to armed combat rescue missions during the Vietnam War.

A number of Seasprites were built as HH–2Ds, similar to the HH–2C but without armour and armament, and some of these were adapted for an anti-ship missile programme, using missiles or Mk. 44 or Mk. 46 homing torpedoes, essentially in an attempt to provide American warships with defence against missile-armed gunboats and fast craft. Existing aircraft were converted after 1972 in considerable numbers to provide an interim helicopter under the LAMPS programme (Light Airborne Multipurpose System), being equipped with search radar, magnetic anomaly detectors (MAD) and sonobuoys, and many of these were attached to American frigates.

The new generation of helicopters for the smaller helicopter-carrying warships of the 1970s, is perhaps best represented by the Westland Lynx, a British-designed helicopter built jointly by Westland in Britain and Aerospatiale in France, and the third member of the trio of helicopters featured in the Anglo-French Helicopter Programme of the late 1960s.

No less than thirteen Lynx prototypes were built including both small-ships naval anti-submarine versions and army utility versions for both the British and French armed services, with the basic concept of

A LAMPS version of the Seasprite prepares to land aboard the escort, USS *Harold E. Holt* in the Pacific.

the helicopter being that of a small helicopter with a crew of two and room for up to ten persons in the cabin immediately behind the cockpit, with either Miniguns, anti-submarine homing torpedoes, HOT or TOW anti-tank missiles, or Sea Skua helicopter-to-ship guided missiles being carried outside the fuselage. As already mentioned French plans for a twin-seat attack helicopter versions were abandoned at an early stage, but one advantage of using what is essentially a utility or light transport helicopter in the attack role, apart from the obvious one of easy role changing, is that additional missiles can be carried inside the cabin, so that the helicopter can reload during a sortie.

Powered by two 900-shp Rolls Royce BS.360–07–26 Gem turbo-shafts, the first flight of a Lynx prototype came in March 1971, and the first few prototypes were basic aircraft. It was not until the fifth prototype that a British Army machine was produced, and the seventh prototype represented a Royal Navy machine. British Army and Royal Navy machines have different avionics and missile control systems, but the most obvious difference is the use of fixed landing skids on the Army machine while the naval versions use a neat retractable undercarriage. The French Navy helicopters have dunking sonar, not fitted on Royal Navy versions, and other detailed equipment differences. A rigid rotor system assists manoeuvrability in the Lynx, which is the first British production helicopter to be able to roll in flight, and a prototype established a European speed record of 200 mph in June 1972, while the range is a useful 390 miles. The first of some eighty machines for the Royal Navy was delivered in 1976, and the first aircraft entered service the following year. The Royal Navy machines have a slightly different role from their Aeronavale counterparts, using the British Aerospace Sea Skua anti-ship missile to provide frigates and destroyers with additional defence against fast missile-armed patrol craft, of the kind supplied in generous quantities to Russian client states. Although the British and French Army machines have been highly satisfactory in service, the Lynx may have lost much of the military market to the Iroquois, and it is as a naval helicopter that this attractive machine has been so successful, with the Royal Navy's eighty machines being followed by orders from the Royal Netherlands Navy, the Royal Danish Navy, the Royal Norwegian Navy, the Federal German Navy, the Brazilian and Argentinian Navies, and a small number for other countries, including Qatar.

The absence of dunking sonar, and its heavy weight of equipment, on smaller helicopters can now be countered to some extent by use of equipment such as Jezebel. Jezebel is a passive sonar system, in that it listens rather than transmits a signal with which to detect submarines, and on larger helicopters such as the Sea King, Jezebel is now available allowing the helicopter to detect the general presence of a submarine without revealing itself, and using the dunking sonar to finally pinpoint the submarine. On smaller helicopters such as the Lynx, a special lightweight version of Jezebel detects the submarine, but either the mother ship's sonar or the helicopter's homing torpedoes have to complete the job. The other drawback of Jezebel is that, while lighter than the dunking sonar, it does tend to be expensive, since the system uses a number of sonobuoys to listen for the submarine and transmit information back to the helicopter, and these are lost after use, while of course dunking sonar can last as long as the helicopter itself or any other part of its equipment.

A later generation of warships able to operate helicopters started with the Royal Navy's Amazon class of eight guided missile frigates of 3,250 tons full load displacement, commissioned between 1974 and 1978. Designed to operate a helicopter such as the Lynx, these fast light frigates are fitted with both Seacat surface-to-air missiles and

189

Strictly for good sailors, a Royal Navy Westland Lynx with torpedoes, aboard the Type 42 guided missile destroyer, HMS *Birmingham*.

Exocet surface-to-surface missiles, and use only gas-turbine propulsion. There is a hangar for the helicopter. The Amazon propulsion system of twin Rolls-Royce Tyne gas turbines for cruising and two Rolls-Royce Olympus gas turbines for high-speed operation is also used by the larger Type 42 Sheffield-class guided missile destroyers, of 3,500 tons displacement, introduced from 1975 onwards, and which also have a hangar for a Lynx helicopter and use the new British Aerospace Sea Dart surface-to-air and anti-ship missile system; some twelve Sheffields are in service or under construction. Another class using the same propulsive system and the same helicopter is the Type 22 Broadsword class, intended originally to act as the successor to the Leander-class frigates, although this duty will now fall to the projected Type 23 frigate, intended to be smaller and cheaper than the Broadsword class, of which some eight vessels are in service, or under construction: this is the first all-missile Royal Navy class, with BAe Sea Wolf anti-missile and anti-aircraft missiles, and Exocet anti-ship missiles, but no main gun armament. No less important, the class can accommodate not one but two Lynx

190

helicopters, providing a rather better chance of having a helicopter available or actually in the air at any time.

The Dutch interpretation of the medium-sized destroyer or frigate is the so-called standard (Standaart) frigate, also of 3,500 tons and intended to consist of the *Kortenaer*, commissioned in 1978, and eventually eleven sisters. These vessels use the same propulsive systems as the Royal Navy ships, and the same Lynx helicopters, with two per ship in wartime, although normally just one will be embarked during peacetime. This class complements rather than replaces the Van Speijk class, which is being modernized in a rather different form from the Royal Navy's Leander class. The Kortenaer class is also being used by the Royal Hellenic Navy's modernization programme, although it remains to be seen whether or not Lynx helicopters will be chosen. The German Type 122 class is based on the same hull form as the *Kortenaer*, itself a development of the Leander or Van Speijk series, but uses different missile systems and a combined diesel and gas-turbine propulsive system known as CODAG, with an American gas turbine. All of the new Dutch warships now use American rather than British missiles. The large destroyer leader Tromp class of two 5,400-ton standard displacement vessels, also uses the Rolls-Royce gas turbine combination of the smaller vessels, and can accommodate a single helicopter of Lynx or Wasp size.

The Italian Navy showed considerable resourcefulness in building major warships able to accommodate large numbers of helicopters for their size, with the first such class being the light cruisers of the Andrea Doria class, both of which were commissioned in 1964 and, despite being of just 6,500-tons full-load displacement, could each accommodate four Agusta-Bell 212 helicopters, in essence a stretched version of the Bell 205, with twin turbo-shaft propulsion.

Even better was to follow, with the *Vittorio Veneto*, a cruiser of 8,850-tons full-load displacement, commissioned in 1969, and which can accommodate up to nine AB.212s. Unfortunately a projected second vessel in this handsome class was cancelled.

The first Italian destroyers to be designed to operate helicopters were the Audace class, just two steam-turbine destroyers, commissioned in 1972 and 1973, which could take two anti-submarine helicopters of AB.212 size, while the two ships of the older Impavido class, commissioned in 1963 and 1964, were converted to take a single helicopter on a landing platform without a hangar during the late 1960s. The small Lupo-class frigates, of which four were commissioned between 1977 and 1980, and which use combined diesel and gas-turbine propulsion, in spite of being of just 2,500-tons full

An Italian Agusta-Bell
204B.

The Italian helicopter
cruiser, *Vittorio
Veneto*.

load displacement, can take two helicopters, although the hangar can provide shelter for just one of these. This successful class of vessel has also been exported to Peru, and a 'stretched' version, the Maestral class, of 3,040 tons, is entering service at present, with the first of eight vessels commissioned in 1981 and the remainder following at intervals through to 1984; Maestrals can take two AB.212 ASW helicopters, and in common with the smaller Italian vessels, use Otomat ship-to-ship missiles and Seasparrow surface-to-air missiles. Helicopters are also accommodated aboard the two Alpino-class frigates, of 2,700 tons, commissioned in 1968.

The French Navy has devoted rather less effort to placing helicopters aboard its ships, due no doubt to the presence of two light fleet aircraft-carriers to lead task forces and provide full air support. Possibly the most interesting French helicopter-carrying vessel is the cruiser *Jeanne D'Arc*, originally laid down as *La Resolue*, but a change of name was permitted after the withdrawal of an older cruiser. The 12,365-ton full-load displacement *Jeanne D'Arc* can accommodate up to eight Westland-Aerospatiale Lynx anti-submarine helicopters, which is far fewer helicopters for her tonnage than the *Vittorio Veneto*. The *Jeanne D'Arc* is older, having been commissioned in 1964, and this may explain the poorer use of space, but possibly more relevant is the need for substantial accommodation aboard the French vessel since she operates as a cadet training ship during peacetime, and becomes an escort or even an assault ship in wartime. Both classes are fitted with guided missiles, the *Vittorio Veneto* using the Tartar anti-aircraft missile and the *Jeanne D'Arc* using six Exocet ship-to-ship missiles.

Few smaller French vessels carry helicopters. The three Tourville-class 5,745-ton destroyers of 1974–7 can take two Lynx, while the current building programme centres around the C70 type, of which a large number may eventually enter service. Otherwise known as the Georges Leygues class, the C70 currently includes six anti-submarine vessels in service or on order, and two anti-aircraft frigates building, and only the ASW version of these 4,170-ton displacement ships can provide hangarage for two Lynx helicopters, although the AA version will have a landing platform. Clearly, the role of the AA vessels is thought not to require even one helicopter permanently embarked, and possibly the two types of warship are meant to operate together, leaving the AA version simply in need of a landing platform for helicopters employed on communications duties.

One of the more interesting fleets of helicopter-carrying warships is that of the Canadian Armed Forces, whose Maritime Command effectively replaces the former Royal Canadian Navy. The novelty of the Canadian approach is that large Sikorsky Sea King helicopters are operated from and accommodated aboard frigates and destroyers of average size, with some ships even being able to take two of these powerful helicopters, although naturally this is at considerable cost in terms of other forms of ship-board armament, such as guns and missile systems, and anti-submarine mortars or torpedoes.

The oldest of the Canadian helicopter-carrying frigates is the 3,051-ton St Laurent class, commissioned in 1956 and 1957, and which were refitted to take one CHSS–2 Sea King during 1963–6, with the helicopter hangar and landing platform replacing a twin 3-inch gun mounting and an anti-submarine mortar. The later Mackenzie, Restigouche and improved Restigouche classes of the late 1950s and early 1960s cannot embark helicopters, but both the two Annapolis-class frigates, 3,000-ton vessels commissioned in 1964, and the larger Tribal-class destroyers, of which four 4,700-ton vessels were commissioned in 1972 and 1973, can embark Sea Kings, with the frigates taking a single helicopter and the destroyers taking two. A new class is being planned as a mid-1980s replacement for the St Laurent-class, and without a doubt this will be a helicopter-carrying type of warship.

As one might expect, most Canadian fleet supply and replenishment vessels also have accommodation for helicopters, and so too do most of the ice-breakers operated by the Canadian Coast Guard.

In contrast with the Canadians, the United States originally tended to take the helicopter less seriously as a small-ships aeroplane, and again, the availability of many aircraft-carriers of differing sizes may have contributed to the relative neglect of helicopter-carrying escort vessels and cruisers, although it is also noteworthy that the United States Navy has only in recent years made a serious attempt to bring its escorts up to modern standards, with a large frigate- and destroyer-construction programme. But, the Royal Navy started wholesale introduction of warships able to operate helicopters at a time when that service maintained five attack and anti-submarine carriers, including the two large vessels of the Eagle class, and two commando carriers.

The Gearing-class destroyers of World War II vintage were the subject of a Fleet Modernization and Rehabilitation-FRAM-programme during the early 1960s, in the course of which helicopter landing platforms appeared on these vessels. However, it was not until the introduction of the Bronstein class of just two vessels in 1963, that any provision was made for helicopters, and these 2,650-ton vessels were also merely equipped with landing platforms. A more positive attitude eventually emerged in the planning of the Brooke class of six

194

3,426-ton displacement and the Knox class of forty-six, 3,825-ton displacement vessels during 1966–74; these ships were intended to operate an unmanned helicopter, known as DASH, or Drone Anti-Submarine Helicopter, although this advanced concept was cancelled at an early stage and instead these ships operated Bell Iroquois and other small helicopters, with, more recently, the new Sikorsky LAMPS Sea Hawk helicopter. However, the current building programme for forty-six Oliver Hazard Perry-class anti-submarine frigates, single-screw gas-turbine vessels of 3,605-tons displacement apiece, will change the picture of helicopter operations from American warships considerably, with each of these vessels able to take two Sea Hawk helicopters and employ these on a variety of anti-submarine and anti-shipping roles. It will also change the image of the Royal Australian Navy as well, since that service is receiving four of these ships, while some may also enter Spanish Navy service, although the Spanish already operate helicopters from their Knox-class frigates.

Although some American destroyers also have helicopter landing platforms, the biggest step forward in helicopter operation from ships of this type in the United States Navy has been with the introduction of the Spruance class, first commissioned in 1975 with the last of thirty-one 7,800-ton displacement guided missile destroyers completing in 1983. Each Spruance-class vessel can take either one HSS–2 Sea King or two Sea Hawk helicopters, and these are being joined by a handful of Kidd-class destroyers, basically an enlarged Spruance originally laid down to an Iranian order before the revolution in that country, and now being pressed into USN service.

Most American cruisers, including the Belknap, Leahy, Bainbridge and Long Beach classes, can now handle helicopters, often after post-

One of the latest American frigates, the uss *Oliver Hazard Perry*. Australia is also using this type of warship.

construction modification, and usually with only a landing platform. This reflects the role of many cruisers as carrier escorts, and therefore the helicopter is a communications and utility machine rather than an embarked anti-submarine weapon.

A number of United States Coast Guard service cutters have been modified to accommodate helicopters, including the relatively large Hamilton and Hero classes, of 3,050 tons, commissioned between 1967 and 1972, which can each take a single HH–3. The six elderly Campbell-class vessels are without helicopters, but the Bear class introduced from 1981 onwards can take a single HH–3 or a Sea Hawk on a ship of just 1,780 tons, and the sixteen Reliance-class vessels, of just over a thousand-tons displacement, commissioned in 1968 and 1969, can also take an HH–3, for which, however, they do not have a hangar.

In common with the Italian Navy, the Soviet Navy has impressed with its penchant for large helicopter-carrying cruisers, including the two vessels of the Moskva class, commissioned in 1967 and 1968, and each of which can carry up to twenty 'Hormone' helicopters on a full-load displacement of 20,000 tons. The more recent Kirov class of battle-cruisers, completing between 1981 and 1984, can take between three and five 'Hormone' helicopters, while cruisers completed between 1967 and the present, including the Kresta I, Kresta II, Kara, Sovremenry and Udaloy classes, can usually handle a single helicopter, even if they cannot always provide hangarage.

The standard Soviet ship-board helicopter is the 'Hormone', the NATO designation for the Kamov Ka–25, another helicopter in the traditional Kamov-mould of tubby contra-rotating rotor helicopters. The Ka–25 was first noticed by Western observers in 1961, when it was designated 'Harp' and appeared with dummy anti-shipping missiles. Eventually production versions with twin 900-shp Glushenkov GTD–3 turbo-shafts and a maximum speed of 130 mph, entered service aboard Soviet warships. While a transport version exists which can carry up to twelve passengers, most of the embarked 'Hormones' are either 'Hormone A', with a chin-mounted search radar, dipping sonar and anti-submarine weapons, or 'Hormone B', with electronics for mid-course guidance of ship-to-ship guided missiles. Distinguishing features of the Ka–25 are the triple-fin tailplane and, unusually for helicopters, an internal weapons bay on the 'Hormone A'.

It may be that a larger helicopter is needed, since recently Mil Mi–14s, developments of the Mil Mi–8, have been introduced in small numbers, generally operating from shore bases.

196

The most recent innovation in the Soviet Navy has been the construction of four Kiev-class anti-submarine cruisers, with an extensive missile armament, an angled flight-deck, and a tonnage of 42,000 tons, which puts them in the medium-sized aircraft-carrier category. First commissioned in 1975, these impressive vessels can usually take a squadron of twelve 'Forger' vertical take-off aircraft, which unlike the British Harrier cannot operate on a short take-off, sixteen anti-submarine 'Hormone A' helicopters and four missile-guidance 'Hormone B' helicopters.

The *Kiev* is typical not so much of the growing interest in aircraft-carriers of the Soviet Navy, which involves the construction of at least one nuclear-powered aircraft-carrier, but of the trend towards special ships for vertical take-off aircraft operation by the world's navies. In some cases, as with the British Invincible class of three 19,500-ton "through deck cruisers", able to operate up to nine Westland Sea Kings, five British Aerospace Sea Harriers and two Westland Lynx helicopters, these ships are a lower cost alternative to the traditional aircraft carrier. In other cases, as with the *Kiev* and the Italian *Giuseppe Garibaldi*, 13,000 tons and intended to operate helicopters only initially, these ships are an extension of earlier, smaller aircraft-carrying vessels, intended to operate helicopters and provide the

A Russian Kamov Ka–25 'Hormone' shipborne helicopter.

Contrast in style and size! The Russian aircraft-carrier *Kiev* is shadowed by a British Leander-class frigate, HMS *Danae*.

An artist's impression of the new Italian aircraft-carrier, *Giuseppe Garibaldi*.

option of other vertical take-off aircraft should these be considered necessary in the future. The United States Navy has considered this option, with the Sea Control Ship design, although none have been ordered, even though this remains an option against the crippling cost of large nuclear-powered carriers, and the possible need to replace many of the older American anti-submarine carriers.

The French, on the other hand, appear to be sticking to the light fleet carrier with conventional aircraft handling, but with nuclear propulsion, for the late 1980s to replace *Clemenceau* and *Foch*.

The Australians are likely to purchase an Invincible class, to replace the ageing light fleet carrier, HMAS *Melbourne*.

Clearly, the helicopter has played an important part in changing the image of the traditional warship. The Japanese, Indian, Brazilian, Argentinian, South African, New Zealand, Chilean and Greek Navies are amongst those currently operating helicopters from warships, extending their capabilities against a growing sophistication particularly amongst the world's submarines. Indeed, the helicopter able to operate from small warships has obviously arrived, just in time.

199

9
ON DUTY TODAY

During the night of Monday, 13th, and Tuesday, 14th August 1979, heavy gales struck the Fastnet Yacht Race, wreaking havoc amongst the 330 yachts of all types and sizes, most of which had already turned at the half-way mark, the Fastnet Rock off the coast of Southern Ireland, and were on the homeward leg sailing towards the finishing line at Plymouth. A long open sea race, open for much of the way to the worst which the North Atlantic can offer, the Fastnet Race has always been regarded as a test of supreme skill for yacht crews, but on this occasion it was to become a test of skill, dedication and courage for the crews of Royal Navy search and rescue helicopters in one of the most dramatic air-sea rescue operations to date.

The first helicopter took off from the Royal Naval Air Station at Culdrose in Cornwall at 06.30 on 14th August, by which time the gale had moderated to a 'breezy' Force 9, with a mean wind speed of 50 knots! At this hour, there were already indications that some twenty-five yachts had foundered or were missing. The first helicopter was a Westland Sea King anti-submarine helicopter of Flight 77, No. 706 Squadron, Royal Navy, one of two Sea Kings standing in as holiday reliefs while the air station, normally used for helicopter training, and the personnel of the resident squadron, No. 771 with its Westland Wessex SAR flight, were on summer holiday. During that long day, two more Sea Kings were flown eight hundred miles from their base at Prestwick in Scotland to Culdrose, away from their normal ASW operations in defence of Royal Navy and United States Navy ballistic submarine bases, to join the rescue. Ground- and air-crew of No. 771 Squadron holidaying near to the base also reported for duty, getting the three SAR flight Wessex helicopters into the air while two Westland Lynx helicopters at the air station, normally used for conversion training, were also pressed into service. Indeed, within a few hours, some twenty aircraft, including Royal Air Force British Aerospace Nimrod maritime-reconnaissance aircraft, from No. 42 Squadron at RAF St Mawgan, near Newquay on the northern coast of Cornwall, were involved in what was eventually to become a four-day rescue effort, with seven British and Dutch warships, including the first of the Type 22 Broadsword class, HMS *Broadsword* herself, on her 'work-up' before joining the Fleet.

Aboard the first helicopter were Lieutenant-Commander Wingate, Lieutenant Simpson and the Leading Aircrewman, later Midshipman, Lapthorn, as pilot, observer and winchman respectively. Flying for a total of fourteen hours, Lieutenant-Commander Wingate and his crew rescued twelve people in appalling weather conditions. At one time, Lapthorn became so exhausted by repeated rescue attempts during which he was frequently completely submerged by waves that Lieutenant Simpson relieved him, although Lapthorn volunteered to remain on duty despite his exhaustion, and flew a total of more than fifteen hours throughout the operation, even though he had had little previous air-crew experience.

On another helicopter, a Wessex, Lieutenant Grayson and Acting Petty Officer Aircrewman Grinney were patrolling close to the Scilly Isles, when, on their second sortie of the day, a Nimrod directed them to a yacht which had lost its rudder. However, the rescue necessitated the helicopter hovering just ten to fifteen feet above the survivors despite forty-foot waves which threatened to engulf the aircraft; small wonder that the *London Gazette* in announcing an award for bravery, was to say that "Throughout this exacting period of flying he displayed great competence and ingenuity of the highest order". Grinney, who was a diver attached to the Fleet Air Arm, was lowered five times into forty-foot seas on this sortie, and on another he had to swim through the stormy seas to reach a life raft, which, with the help of his helicopter, he then towed clear of a yacht whose dangerously swinging mast threatened the helicopter and prevented a more conventional rescue.

Other helicopters acted in a similar manner, while others, or often the same machines and same crews flying up to ten hours out of a thirteen-hour period, plucked survivors from dangerously spinning life rafts while their aircraft was low on fuel. That the operation could take place at all was a tribute to the development of the helicopter from a frail machine with limited prospects during inclement weather to one capable of facing the severest demands. Described by a Royal Navy photographer as being "like a battlefield, with yachts all over the place", the pace of the rescue and the size of the race made it difficult until relatively late in the operation to identify those yachts which had had their crews rescued and others which were awaiting assistance or whose crews had been washed overboard. Eventually, however, the abandoned yachts were gathered together, enabling the helicopter crews to see which vessels still needed assistance.

There were the inevitable problems. Working in any aeroplane at low altitude with the door open to the elements is uncomfortable. One

Royal Navy Surgeon Lieutenant, David Morgan, told newspaper reporters that working in the back of a Sea King helicopter was cold, noisy, dark, damp or even soaking wet and with vibration and a working space of just four foot by six foot, so that it generally resembled resuscitating a patient in a cramped bathroom in the dark! Static, accumulated by the helicopters during their flights, ran to earth as winchmen landed on the decks of yachts, giving them what were often severe electric shocks, and adding to the problems of weather, high seas, exhaustion and dangerously swaying rigging and masts. Journalists, while being given a pre-flight briefing before being taken over the scene, were warned that they should not inflate their life-jackets before leaving the helicopter in the event of a ditching for fear of being trapped inside, and that they should check that the rotor blades had stopped turning before leaving, since these continued to rotate under water. Once clear of the helicopter, when rescue arrived, they were told not to grab hold of the rescue strop as it was lowered, "Otherwise,"—as one Royal Navy man put it, "we will be lowering a coffin."

In spite of the difficulties, over four days, nine helicopters flew for a total of some 200 hours, flying pre-dawn to 20.00. Some eighty yachts were lost, almost a quarter of those taking part, and the Royal Navy helicopters accounted for seventy-nine survivors; a total of seventeen yachtsmen were lost. Eight Royal Navy air-crew received the Queen's Commendation for Valuable Service in the Air, including Lieutenant-Commanders Brock and Wingate, Lieutenants Fox, Grayson and Simpson, Midshipman Lapthorn, Acting Petty Officer Aircrewman Grinney and Leading Aircrewman Burnett. There were other awards too, including gold medals presented at the following year's London International Boat Show, and a special dinner held in honour of the naval air-crew by the Yacht Club de France.

A major disaster in 1980 to the Norwegian-owned oil rig, Alexander Kielland, also saw SAR helicopters in action, including one of the RAF's then recently delivered Westland Sea King HAR3s, which rescued ten survivors and directed rescue vessels to twenty-six others.

However, the rescue services have not been without their own tragedies, and although extremely rare in spite of operating in severe weather conditions and often with crews working at the limit of exhaustion, one of these accidents was unusual. On 18th November 1980, another of the RAF's Sea Kings, from No. 202 Squadron who provided the SAR flight at Coltishall in Norfolk, was called out to rescue the pilot of a United States Air Force A–10 attack aircraft operating from RAF Bentwaters in Suffolk; the aircraft had crashed into the North Sea following a mid-air collision with another A–10. Master Air Loadmaster David Bullock saw the American pilot floating, apparently unconscious, in the water, but trapped in the cords of his parachute and being pulled seawards by a strong tide. Bullock was lowered into the sea to pick up the American airman, but he too became trapped in the parachute cords, and when his helicopter tried to lift him, the winchline snapped, leaving Bullock in the sea. The main problem, and the reason for the snapped line, was of course, the weight of water in the waterlogged parachute of the USAF pilot. A USAF Sikorsky HH–53 Super Jolly Green Giant from RAF Woodbridge quickly joined the rescue operation, being accustomed to working closely with their RAF colleagues, but on recovery, both the USAF pilot and Master Air Loadmaster Bullock were found to have drowned.

Such an accident clearly demonstrates the difficulties which helicopter crews have to face, even on what could have been a fairly routine operation. There were suggestions subsequently that the Royal Navy system of using winchmen trained as divers could have prevented the double tragedy, but even a diver can become entangled in parachute cords. Perhaps the tragedy is best put into perspective by the fact that during the year which ended with this sad loss, No. 202 Squadron's helicopters had rescued 254 civilians and 28 servicemen.

Helicopters continue to play an important role in all forms of disaster relief. When Mount St Helens, a volcano thought to be extinct, erupted in 1980, helicopters provided the main form of relief and communications throughout those parts of the north-west United States covered in heavy deposits of volcanic ash. When that same year, an Air New Zealand McDonnell Douglas DC–10 crashed on the slopes of Mount Erubus in Antarctica, only helicopters could move accident investigation teams to the wreckage. Even in less extreme conditions, the helicopter often finds itself at the forefront of such gruesome duties, for example French Army Alouette IIIs, used to transport investigators to the scene of an Inex-Adria DC–9 crash in Corsica in December 1981, and to help in moving the bodies from the wreckage.

The rescue, or perhaps to be more precise, the recovery, of American astronauts after their splash-down in the Pacific Ocean on return from space missions, must be one of the better publicized of all the helicopter's many rescue efforts. The use of a United States Navy helicopter from a carrier, including on some occasions Marine carriers or amphibious assault ships, became standard practice and continued throughout the Apollo moon-landing programme. Navy frogmen would be dropped to fasten a flotation collar around the space capsule, before the astronauts were allowed to open their escape hatch to be picked up by a helicopter, usually a Sikorsky SH–3D Sea King. Typical of these missions was the return of the Apollo XI command module *Columbia* from the first lunar landing on 24th July 1969, when Flight Commander Neil Armstrong, who was the first man to walk on the moon's surface on 21st July, and Colonel Edwin Aldrin, USAF, and Lieutenant-Colonel Michael Collins, USAF, were rescued by a SH–3D of Squadron HS–4 from the aircraft-carrier, USS *Hornet*.

However, the bulk of the world's helicopters are in military and naval service, and of these most are assigned to operations which have nothing to do with humanitarian motives. The helicopter is now the key element in the defence of most nations from enemy tank or submarine attack, and in the fast transport of troops into battle. Because of the cost of buying, maintaining and operating helicopters, and the demand for exceptionally skilled air-crew and ground personnel, the helicopter is most prevalent in the developed nations, and is still less in evidence in the developing or Third World nations.

The world's largest fleet of helicopters throughout the entire history of this type of aircraft has been, and remains, that of the United States Army, which today operates some 9,000 machines, compared to just 1,000 fixed-wing aircraft. Almost half of the helicopters in the US Army are Bell UH–1 Iroquois, operating in the tactical transport, CASEVAC and attack roles, and supported in the liaison and communications role by more than 2,000 Bell OH–58A Kiowas, the military version of the highly successful and popular Bell 206 JetRanger, and by some 1,400 Hughes OH–6A Cayuse light observation helicopters. The core of the attack helicopter force is 1,000 Bell AH–1G and AH–1Qs HueyCobra, operating with TOW anti-tank missiles and with the earlier versions being gradually replaced by some 700 AH–1S Cobras. Heavier battlefield support and anti-tank attack is provided by the first of 350 Sikorsky UH–60A Black Hawks and the first of a planned 536 Hughes AH–64 advanced attack helicopters. Heavy lift and the mainstay of the battlefield mobility is provided by 456 Boeing-Vertol CH–47A/B/C Chinooks and eighty Sikorsky CH–54 Tarhe flying-crane helicopters. The massive training support of this large force requires some 300 Hughes TH–55A helicopter trainers, in addition to some elementary fixed-wing training, while some older types of helicopter remain on base and second line duties.

Frequently operating in the same battlefield as the United States Army, and also often being positioned by the United States Navy for an assault on an enemy beach, the United States Marine Corps tends to place the main emphasis on the assault role. Mainstay of the assault squadrons are a force of some 270 Boeing-Vertol CH–46E/F Sea Knights, supplemented by a heavy-lift force of about a hundred Sikorsky CH–53D/F Sea Stallions and a small number of the triple-engined Super Sea Stallions. A utility force includes more than 120 Bell UH–1E/N Iroquois and a helicopter attack force uses about eighty Bell AH–1J SeaCobra fast-attack helicopters, an improvement over the early HueyCobra, and which are usually armed with either TOW anti-tank missiles or unguided rockets. Generally, a number of these types are operated from the assault ships, including the Tarawa class, which can accommodate up to thirty helicopters at any one time, sometimes with, or instead of, vertical and short take-off and landing aircraft such as the British Aerospace AV–8A Harrier, and the joint British Aerospace-McDonnell Douglas AV–8B development.

The United States Navy naturally remains as a major helicopter operator in its own right, including no less than thirteen squadrons of Sikorsky SH–3D/H Sea King anti-submarine helicopters, which are frequently embarked aboard the USN's fleet of aircraft-carriers. The new classes of destroyers and frigates are receiving the bulk of the fleet of more than two hundred Sikorsky SH–60B Sea Hawk LAMPS helicopters, some of which will replace the Sea Kings aboard the aircraft-carriers in due course, and naturally the initial orders may be increased. The USN also operates eighteen Sikorsky CH–53E heavy-lift helicopters, while thirty of the basically similar RH–53Ds are operated in three mine counter-measures squadrons, and there are also four squadrons of Boeing-Vertol UH–46 Sea Knights and Sikorsky HH–2Ds. There are no less than seven USN reserve squadrons, operating helicopters, including earlier versions of the SH–3, while helicopter training is with TH–1s, HH–46s and UH–1Hs and TH–57As.

Although operating the smallest of the American service helicopter fleets, the United States Air Force is a major helicopter operator in its own right, using helicopters in the rescue role, for missile-site support and for transport and communications duties. One of the key elements of the USAF helicopter fleet in the future will be the Sikorsky UH–60, relative of the US Army's Black Hawk and the USN's Sea Hawk, which will not only update the force but also introduce greater standardization by replacing a mixture of Bell UH–1s, Sikorsky CH–3s and CH–53s, and perhaps later the Military Airlift Command's force of 120 UH–1Fs, on missile-site support, and thirty UH–1Hs on base rescue duties. The rescue helicopter force is generally under the control of Military Airlift Command, which operates fifty Sikorsky HH–1Ns and forty-five HH–3Es, as well as half of the USAF's total of more than seventy HH–53B/Cs, or Super Jolly Green Giants, many of which are operated by USAF Europe.

The United States Coast Guard Service operates a helicopter fleet of Sikorsky HH–3s and HH–2s, with the latter being replaced by a small number of HH–60s, while as a lightweight rescue and communications helicopter, the USCG has recently introduced the first of sixty Aerospatiale Dauphins.

A neighbour of the United States and NATO ally, Canada does not maintain separate armed forces, instead operating a unified defence force, the Canadian Armed Forces, which is divided into groups which approximate to naval, army and air defence formations. The country's lack of commitment to defence ensures that the Canadian Armed Forces are far below the size which one might expect from a country of Canada's size, population and wealth, maintaining armed forces which suffer from a shortage of first-rate equipment. Of the Canadian Armed Forces commands, Tactical Air Command is the one charged with support of the ground forces, and for this it uses some fifty Bell CH–135 Hueys and more than sixty Bell CH–136 Kiowas, both mainly on transport and liaison roles while large Boeing CH–47 Chinooks, but only eight of these, provide heavy transport support. Eleven of the Kiowas operate with the Canadian Army Corps in West Germany. Maritime Command, which operates ships as well as aircraft, has thirty-five Sikorsky CH–124 Sea Kings, most of which were assembled in Canada, for ASW duties, including operations from escort vessels and replenishment vessels.

With the other NATO allies, the size of their helicopter element varies according to national wealth and the size of the armed forces concerned. The Belgian Army, for example, operates a small force of Aerospatiale Alouette IIs on liaison and aerial observation post duties, alongside Belgian-built Britten-Norman Islander or Defender fixed-wing aircraft on utility and liaison duties. The Belgian Air Force, the FAB, operates just five Westland Sea King Mk. 48s along the narrow Belgian coastline for SAR work, although one is normally held in reserve as a VIP transport. The Belgian Navy has just three Alouette III helicopters, mainly for liaison duties since the Belgian Navy is primarily committed to mine counter-measures as part of a specific specialization in the assignment of duties within NATO.

A rather larger naval air arm is operated by the Dutch Navy, the Marine Luchtvaartdienste, or Royal Netherlands Navy Fleet Air Arm, with twenty-four Westland Lynx helicopters used on SAR, ASW and VIP duties, including helicopters assigned to the vessels of the Kortenaer and Van Speijk classes. The Lynx helicopters replaced a smaller number of Westland Wasps, sold to India during the early 1980s, while a larger helicopter force was required to meet the growth in the number of helicopter-carrying vessels, and the fact that the Kortenaer class take up to two helicopters. The Royal Netherlands Air Force is another Alouette III operator, with seventy-two, and also thirty MBB BO.105Cs, all of which are operated on behalf of the Dutch Army, which in practice has no aircraft of its own.

Further north, the Royal Danish Navy operates eight reconnaissance and anti-submarine Westland Lynx from its newer frigates, with a similar number of Alouette IIIs on liaison duties, while the Danish Army operates fifteen Hughes 500M on AOP duties. A small shore-based SAR squadron is operated by the Royal Danish Air Force, using eight Westland S–61A Sea King helicopters.

A similar pattern is repeated in Norway, which has six Westland Lynx operating from coastguard vessels on SAR and liaison duties, while the Royal Norwegian Air Force operates ten search and rescue Westland Sea King Mk. 43s as well as twenty-two Bell UH–1B Iroquois helicopters, the latter being assigned to transport and army support roles.

A fairly large Westland Sea King operator is the West German Marineflieger, with twenty-one Mk. 41 helicopters on SAR duties, while twelve Lynx were delivered in 1981 for use from the six Type 122 frigates of the West German Navy. A large number of Bell UH–1D

A Royal Danish Air Force S–61.

A line-up of Royal Norwegian Air Force SAR Sea Kings.

utility helicopters is operated by the Luftwaffe, which has about a hundred of these machines at its disposal. However, Western Europe's largest helicopter operator is the Heeresflieger, or Federal German Army Air Corps, which has some seven hundred helicopters, as befits the best-equipped West European army. The main type, as one would expect is the German MBB BO.105M, of which two hundred operate on liaison and communications duties, while another two hundred Dornier-built Bell UH–1Ds operate on utility and light transport duties, backed by 108 VFW-built Sikorsky CH–53Gs on heavy transport and assault duties, while most recently this well-equipped force has been given some teeth, with the arrival of two hundred BO.105Ps, equipped with HOT anti-tank guided missiles. Each brigade in the German Army has its own Army Aviation Command element.

Although outside NATO's command structure, France is still nominally a member of the Alliance, with a strong and well-equipped air force and reasonably large and adequately equipped ground and naval forces, although the army has been relatively neglected in recent years, particularly compared to the Armée de l'Air, the French Air Force. This said, the Aviation Légère de l'Armée de Terre is strong, with some six hundred helicopters, mainly Aerospatiale Alouettes, Aerospatiale-Westland Gazelles and Pumas, with the last-mentioned providing a substantial medium-lift capability, with about 135

A German Army MBB BO.105P fires its HOT anti-tank missiles.

Another look at an BO.105P, on the ground.

machines in service. Almost two hundred Alouette IIs are used in the light observation role, while seventy Alouette IIIs are equipped with AS.11 anti-tank missiles and operate alongside some forty HOT-equipped Gazelles and 128 HOT-equipped SA.342Ms, which are usually assumed to be available to NATO in the event of a Russian attack on the Central Front, in contrast to the doctrine during the Gaullist era, when France claimed that she would only attack if attacked first, and Russian troops entered her own territory.

The Aeronavale operates forty Aerospatiale-Westland Lynx on anti-submarine duties, including a number of aircraft embarked aboard frigates, supplemented by a number of liaison Alouette IIIs, and ten each of Aerospatiale Super Frelons in both the assault and anti-submarine roles aboard the aircraft-carriers *Clemenceau* and *Foch*. A small number of helicopters is operated by the French Air Force on VIP duties.

In Britain, the Army is also the country's largest helicopter operator, with about four hundred aircraft, including some two hundred Westland-Aerospatiale Gazelles, on AOP and communications duties or, with TOW anti-tank missiles, on anti-tank duties, as well as more than a hundred Westland Lynx helicopters, usually equipped with TOW anti-tank missiles and also available for light transport duties, while there are also a few remaining Westland Scout helicopters and a small number of Alouette IIs in Cyprus. Many of the British Army's helicopters are based with the British Army of the Rhine in West Germany, although others are operated in Northern Ireland, but this is not the scene of any massive helicopter deployment, with perhaps no more than twenty Army helicopters in the province at any one time. Army Air Corps training is on Bristow-operated Bell 47Gs, although most of these have now been replaced by Gazelles.

One of the British Army's new TOW–equipped Westland Lynx helicopters.

The Royal Navy is the largest helicopter operator of any European navy, with some seventy Westland Sea King Mk. 1 and Mk. 2 helicopters operating in the anti-submarine warfare role, with these helicopters assigned to shore bases and to the aircraft-carriers, HMS *Hermes, Invincible, Illustrious* and *Ark Royal*, of which the last will enter service in 1985. HMS *Hermes* is likely to be retired during 1983, but HMS *Invincible* is no longer to be sold to the Royal Australian Navy. A further seventeen helicopters of the Westland Commando version, a development of the Sea King, known in British service as the Sea King HC4, are operated by the Royal Marines, usually with Royal Marine pilots although Royal Navy pilots are also used. Small-ships helicopters include eighty-eight Westland Lynx operating from all of the Royal Navy's frigate and destroyer classes on both anti-submarine and, with the new Sea Skua anti-ship missile, on anti-shipping duties, while a small number of Westland Wasps remains, and there are some SAR Westland Wessex, as well as some of the ASW version of this helicopter operating from the remaining County-class destroyers.

Another major helicopter operator, the Royal Air Force operates helicopters in the transport and SAR roles, as well as a few on communications duties. The SAR element has recently been modernized with the replacing of elderly piston-engined Westland Whirlwinds by sixteen Westland Sea King HAR3s, while a number of Westland Wessex also operate in this role. Some Wessex helicopters still operate on transport duties, but the mainstay of the transport force today is the Westland-Aerospatiale Puma, of which there are more than forty, while recently thirty-three Boeing-Vertol CH–47D Chinooks have entered service in the heavy-lift role. Communications helicopters are mainly Westland-Aerospatiale Gazelles, and this type is also the main RAF training helicopter.

On NATO's southern flank, the Italian Army's Aviazione Leggera del l'Esercito (ALE) is a major helicopter operator, including its own heavy-lift element with twenty-four Meridionali-built CH–47C Chinooks. Smaller helicopters include 140 Agusta-Bell AB.206 Kiowas, 130 Agusta-Bell AB.205s and 40 AB.204 Iroquois operating in the attack and light transport roles, while liaison and AOP roles are performed by the fleet of about fifty AB.47G/Js. The new Agusta A.109 Hirundo is being operated on a trial basis with TOW missiles for anti-tank duties, and it is believed that at least sixty will enter service if the trials are successful, while Europe's first production battlefield attack helicopter, the A.129 Mangusta, is also being evaluated in the anti-tank role.

In recent years, the Marinavia, the Italian Navy's air arm, has

211

enjoyed a rapid expansion, building up a force of thirty Agusta-built SH-3D Sea Kings, up to sixteen of which can be operated from the helicopter carrier *Giuseppe Garibaldi*, scheduled to enter service in 1983, while for operation from smaller ships there are almost fifty Agusta-Bell AB.212 helicopters, which replaced a large number of AB.204ASs on anti-submarine duties, and all of which are equipped with dunking sonar and can carry homing torpedoes. The AB.204B remains in service with the air force, the AMI-Aeronautica Militare Italiano, which operates almost fifty on liaison, SAR and VIP duties, alongside a force of fifty Agusta-Bell AB.206s, while twenty Sikorsky HH–3Fs are assigned to SAR operations.

Meridionali-built Chinooks are also in service with the Greek Army, which has six operating in the transport role, with just eight Bell AH–

A Meridionali-built Boeing CH–46 Chinook shows its flying-crane potential.

1S HueyCobras providing a token anti-tank battlefield element, supporting older Agusta-Bell 204 and 205 helicopters, of which there are fifty. A small force of Bell 47Gs operates on utility, liaison and AOP duties. The Greek Navy's ASW helicopter force is also largely Italian in origin, with sixteen AB.212s, while there are also four Alouette IIIs assigned to liaison duties. It is likely that the AB.212 will also be operated from the first of the Greek Navy's new frigates, which will be the Dutch Kortenaer class, rather than the Westland Lynx.

For many years, the defence forces of both Greece and Turkey suffered from little or no military assistance from the United States, essential for these two relatively poor nations, due to mutual antagonism between these two NATO members over the Cyprus issue and the question of territorial and economic zone rights in the Aegean Sea. The improvement in relations in recent years and the return of Greece to full NATO membership after a period of refusing to play an active role within the alliance, has seen a resumption of American aid. Even so, the Turkish Army, or Turk Kava Kuvvetleri, operates relatively few helicopters for its large manpower size, with just one hundred Agusta-Bell AB.205s in the transport and gunship roles, with another twenty AB.204s and fifteen of the small AB.206s on liaison duties. The Turkish Navy has just three AB.205s on liaison and light transport duties, and six AB.212s on ASW duties. A few older helicopters, including Sikorsky SH–19Bs, are operated by the Turkish Air Force on transport and communications duties.

Portugal's armed forces also suffered from a lack of military aid during the period of dictatorship and a series of African colonial wars. However, recently the Portuguese Army has introduced ten Agusta A.109s on anti-tank duties, and these are the mainstay of the Army's small helicopter force, while the Navy operates few helicopters and the Forca Aerea Portuguesa operates twelve transport and SAR Aerospatiale Pumas, and up to forty liaison and utility Alouette IIIs. During the Salazar regime, it was also difficult for Portugal to purchase military equipment, but in common with Greece and Turkey, this is one of NATO's poorer members.

NATO's newest member, Spain, operates more than one hundred helicopters in the Fuerzas Aeromoviles del Ejercito de Tierra, with twelve Boeing-Vertol CH–47 Chinooks in the heavy transport role and an anti-tank force of sixty MBB BO.105s, with a number of Bell UH–1H Iroquois and Sikorsky S–58s in the transport and utility roles, supplemented by small numbers of Bell 206s and 47s. The Spanish Navy also operates a number of Sikorsky SH–3Ds and Agusta-Bell 212s from frigates and from the aircraft-carrier, *Daedalo*, otherwise the

USS *Cabot*, one of the few surviving Independence-class light fleet carriers of World War II vintage, on loan from the US Navy.

Opposing these western forces, the Soviet Union operates a massive array of air power divided into a number of key independent commands, such as Frontal Aviation, Naval Aviation and Long-Range Aviation, with the first two operating large numbers of helicopters. Frontal Aviation, as the name suggests, is essentially a tactical transport and strike force operating in support of ground forces, with large numbers of Mil Mi–24 'Hind' anti-tank attack helicopters, including the latest 'Hind D', with a Gatling-type machine-gun in a chin turret, and the anti-tank missile-armed 'Hind E', with tactical transport and assault support maintained by the Mil Mi–8 'Hip'. Additional transport helicopter support is provided by more than 2,500 helicopters of all ages and sizes, ranging from the now obsolete Mil Mi–1 and Mi–2 in the liaison and utility roles through the Mi–4s, although the mainstay of the fleet is the Mi–8 and the large Mi–6 and Mi–10 heavy lift helicopters, and some more recent types including the big, new Mi–26 transport helicopter and Mi–17.

Soviet Naval Aviation operates some five hundred helicopters, mainly of Mi–4, Mi–8 and Kamov Ka–25 types, with recent deliveries of large numbers of an anti-submarine variant on the Mi–8 series, known as the Mi–14. While the larger Soviet warships can accommodate up to twenty Ka–25s, these types are used not only for anti-submarine duties, but also for mid-course guidance of ship-to-ship missiles. More than half of all Soviet naval helicopters are believed

The ample carrying capacity of the new Mil Mi–26 heavy transport helicopter is clearly demonstrated here.

to be of the Ka–25 type.

Many of the Soviet satellite states operate helicopters in large quantities, as part of the Warsaw Pact's tactical forces. The great extent of the standardization of equipment to be found in the Warsaw Pact forces means that most of these air arms operate smaller numbers of the helicopters to be found in Soviet service, including, for example, in the case of the Bulgarian Air Force, forty Mil Mi–4 and Mi–8 helicopters, with a quantity of Mi–24 attack helicopters. The Czech Air Force is rather larger, with about one hundred Mi–4s and Mi–8s, and again, a force of Mi–24s, while this pattern is repeated in East Germany, although this country is one of the two Soviet satellites with a naval aviation element, operating a force of naval helicopters in the Baltic, with Mi–4s and some Ka–25s. Oddly, Hungary operates just a few Mi–4s and Mi–8s, some Mi–24s and land-based Ka–26s. A larger helicopter force is maintained by Poland, with some two hundred machines of the main types in the air force and a purely land-based naval helicopter unit with a handful of Mi–4s and Mi–8s for liaison and utility duties.

A Polish-built SM–2 helicopter, used on Army co-operation and communications duties.

The odd one out amongst the Warsaw Pact forces is the Romanian Air Force, which reflects in its equipment some of the independence of outlook of that country's foreign policy; this is the only Pact member not to have Soviet forces based on her territory. A small number of Mi–4s and Mi–8s on transport and search and rescue duties are outnumbered by some fifty locally built Alouette IIIs on anti-tank operations and a large force of ninety licence-built Aerospatiale Pumas for transport and assault duties.

Albania, a non-Warsaw Pact Communist republic, operates a small force of about thirty Mi–1s and Mi–4s in the People's Army Air Force. Another Communist republic, neighbouring Yugoslavia, operates a mixture of Eastern Bloc and Western equipment, reflecting her apparent neutrality, with the mainstay in the Yugoslav Air Force's helicopter units being about a hundred Aerospatiale Gazelle anti-tank helicopters built under licence in Yugoslavia, which have replaced some of the earlier Mi–4s and Alouette IIIs, while a few Westland Whirlwinds and Mi–8s remain, supplemented by a small number of Agusta-Bell AB.205 Iroquois. The Gazelles are fitted with Soviet Sagger anti-tank missiles. The Yugoslav Navy operates a few Gazelles on liaison duties, with a number of Mi–8s and anti-submarine Ka–25s.

Of the non-aligned European nations, Austria operates a small air force, the Oesterreichische Luftstreitkrafte within the Federal Army, with three squadrons operating helicopters, including twenty-two Alouette IIIs on search and rescue and liaison duties, a couple of Sikorsky S–65s on heavy transport duties, and twenty-four Agusta-Bell 212s on transport and utility work, and a dozen Bell 206As and an equal number of Hughes OH–58Bs on AOP duties. The Swiss Air Force operates more than a hundred Alouette II and III helicopters, the sole helicopter element in this relatively strong air arm. Further north, in Sweden, the Swedish Navy has control of all helicopter procurement and training, operating a small force of ten Boeing and Kawasaki-built 107s for both anti-submarine and mine counter-measures operations, with a similar number of Agusta-Bell 206s and a training force of five Alouette IIs. The Swedish Army has twelve Agusta-Bell 204B Iroquois, forty 206Bs and six Alouette IIs. Finland's helicopter force is just a couple of Hughes 500s, and eleven Mi–8s on transport duties.

Outside Europe and North America, Japan is the largest operator of helicopters, with a force of thirty SAR Kawasaki-built Boeing-Vertol 107s and seven Sikorsky S–62As in the Japanese Air Self-Defence Force, while the Japanese Maritime Self-Defence Force operates both S–61s and S–62s on SAR, with an ASW force of more than sixty

216

SH–3A/Bs, and mine counter-measures units with KV–107s. The Japanese Ground Self-Defence Force operates more than sixty Bell AH–1S Cobras, 150 Hughes OH–6J/Ds, 150 Bell UH–1B/H and a small number of KV–107s, which are being replaced by forty Boeing-Vertol Chinooks to provide an enhanced heavy-lift element.

India, by contrast, operates a mixture of Eastern and Western equipment, partly reflecting that country's neutrality and partly the desire to take whichever seems to be the best bargain on offer at any one time. The Indian Air Force operates about a hundred Aerospatiale SA.315 Cheetah helicopters, with about the same number of Mil Mi–4s and about fifty Mi–8s, while a force of 120 Alouette IIs are operated on Army liaison duties. This mixture also occurs in the Indian Navy, which operates twelve ASW Westland Sea Kings, some ex-Dutch Westland Wasps and more than twenty ASW and SAR Alouette IIIs alongside Kamov Ka–25s. The British and French aircraft are operated from the aircraft-carrier *Vikrant*, a British-built CVL, while the Ka–25s operate from Russian-designed Kashin-class vessels and, occasionally, from Leander-class frigates built in India, although these vessels and a new stretched class of Indian Leander normally use French helicopters. The few American helicopters in the Indian armed forces include Hughes 300s, used solely for training helicopter pilots.

China, as one would expect, has a tradition of using licence-built Soviet designs, including the Mil Mi–4, known in China as the H–5, and a relatively small number of Mi–8s, but in recent years a small number of Aerospatiale Super Frelons have also been acquired. However, for the size of the armed forces, helicopters are but a small element, with just 350 machines altogether. Both China and India suffer from maintaining large armed forces with little sophisticated equipment due to foreign-exchange problems and a relatively limited industrial base of their own, while the prevalence of skilled manpower

An Alpine backdrop for an Austrian Aerospatiale Alouette III.

or manpower with the potential to absorb skills is relatively low.

The best equipped of all the air forces in the Middle East is the Israeli Defence Force/Air Force, which operates a small but effective anti-tank helicopter element with some fifty TOW-equipped Bell AH–1G/S/Q helicopters, complemented by a relatively strong assault and transport helicopter force of some eight Aerospatiale Super Frelons and thirty Sikorsky S–65Cs, while utility and light transport duties are performed by fifty Bell 205 Iroquois and 206 Kiowas, and twelve Bell 212s operate on SAR duties. The Egyptian Air Force is also a considerable force, but with a mixture of Eastern and Western aircraft types, reflecting the change in the country's allegiances over recent years. Few Soviet types actually remain in service, with the Mil Mi–4s, Mi–6s and Mi–8s being gradually replaced by some fifteen Agusta-built Boeing CH–47C Chinooks while the Army now has more than fifty Aerospatiale Gazelles, a few of which are loaned to the Egyptian Navy to assist in liaison and communications duties, complementing a small number of Westland Sea Kings, which operate on anti-submarine duties, while the Army itself is building up an effective force of Westland Commando troop-carrying helicopters. Iraq also operates a very international assortment of aircraft types, although if anything the Iraqi regime is moving in the opposite direction to that of Egypt, with a dozen elderly Westland Wessex helicopters, forty or so Alouette IIIs and a similar number of Gazelles, and twenty Pumas, operating alongside thirty-six Mil Mi–4s and Mi–8s, and a small number of Mi–1s, Mi–6s and Mi–10s, in the Air Force, while the Navy operates a small force of Agusta-Bell 212s on ASW duties. Iraq's neighbour and long-time rival, Iran, maintains a force

The Israeli Defence Force also uses anti-tank OH–6As.

which is almost entirely American in origin, including large numbers of Bell 204 and 205 Iroquois light transport helicopters and Meridionali-built Boeing-Vertol Chinooks, but the operational condition of this force is uncertain, following the embargo on military goods and supplies for the Iranian revolutionary regime after the seizure of the American hostages already mentioned earlier in this book.

On the other side of the Persian Gulf, there is rather greater consistency in procurement policy with close ties between these countries and the West. Saudi Arabia, the main political and military power in the region, operates a mixture of American and French helicopter types, including Italian-built Agusta-Bell AB.205s, of which twenty-four operate in the liaison and SAR roles, with sixteen AB.206s on training and SAR duties, alongside a small number of Bell 212s and Alouette IIs, which have recently been joined by twenty-four Aerospatiale Dauphin helicopters armed with the AS.15TT air-to-surface anti-tank missile. The small, but efficient, Kuwait Air Force operates two squadrons, each of twelve Aerospatiale Gazelles, with one squadron operating in the AOP role and the other on anti-tank duties, while transport is provided by a squadron of twelve Aerospatiale-Westland Pumas. At the other end of the region, the Sultan of Oman's small Air Force operates fifteen Agusta Bell 205s and five AB.212s on utility and SAR duties, with three AB.206s. Qatar, in spite of a small population, maintains a dozen Westland Commando transport helicopters and a small number of Pumas, as well as some Westland Widgeons and Aerospatiale Gazelles for operations by the Qatar Emiri Air Force, while the smaller helicopters are also made available to the Qatar Police. In common with many of the air forces in this strategically vital and sensitive region, the Royal Jordanian Air Force has received new equipment in recent years, with the introduction of eight Hughes 500D training helicopters and a small number of both the new Sikorsky S–76 Spirit and Bell AH–1Q attack helicopters, supplementing the mainstay of the force during the 1970s, a squadron of fifteen Aerospatiale Alouette IIIs.

In contrast to these Western-aligned nations, Syria has taken a rather less easily identifiable stance, resulting in the Syrian Arab Air Force maintaining a force of about fifty Mil Mi–8s and about thirty Mi–4s, along with smaller quantities of Mi–6s and even Mi–24 attack helicopters, while the Navy operates nine Kamov Ka–25s on anti-submarine duties, but in recent years the Air Force has also received twenty Aerospatiale Gazelles for anti-tank use and six Meridionali-built CH–47C Chinooks, while the Navy has introduced twelve

219

Agusta-built SH–3D Sea Kings and eighteen AB.212s.

Relatively few of the North African air forces operate helicopters in large numbers, although Libya operates a few machines of Soviet origin, possibly including some Mi–24s, and Tunisia has received some modern American equipment as a result of being involved in border clashes with Libyan forces during 1980. Morocco is the exception, with no less than twenty-four anti-tank Hughes OH–6A, donated by Saudi Arabia, and more than thirty Agusta-Bell AB.205s and AB.206s, as well as a small number of AB.212s, Aerospatiale Alouette IIs and Gazelles, and some forty Pumas on transport duties.

The poverty of many African countries and the sheer shortage of skilled personnel or sufficient educated recruits, limits their ability to operate helicopters in quantity, and indeed many can hardly manage to finance and operate fixed-wing aircraft. There are a few exceptions, however, including oil-rich Nigeria, with a small but modern air force, amongst whose aircraft can be included twenty MBB BO.105s for SAR duties, ten Aerospatiale Pumas for transport work, and a collection of elderly Westland Whirlwinds and Aerospatiale Alouette IIs. Perhaps rather more typical is nearby Ghana, with just four Alouette IIIs and two Bell 212s. As one of the more prosperous countries in East Africa, Kenya operates a modern, but small, force of fifteen Hughes 500MD Defender anti-tank helicopters armed with TOW missiles, in addition to the same number of Hughes 500 'Quiet Armed Scout' AOP helicopters and three Alouette II and two Bell 47G utility helicopters; the more modern helicopters have been supplied by the United States under the Foreign Military Sales Programme. Further south, in

Kenya is one of the few African countries to use helicopters in any number, with a small but potent force of anti-tank Hughes OH–6As.

Angola, the remains of about twenty ex-Portuguese Alouette IIIs abandoned after Portugal granted independence to the country in 1976, are operated by the Angolan Air Force on counter-insurgency duties against pro-Western UNITA and other anti-government troops. It is probable that many of the Angolan helicopters are flown by Cuban pilots, or by other mercenaries, and some may be used in support of the SWAPO guerilla organization in its raids into South-West Africa or Namibia.

South Africa operates a large number of helicopters, mainly of French origin due to sanctions by both the British and American administrations of recent years. It is believed that as many as sixty Pumas may be operated in the tactical transport role, with a medium-lift capability provided by up to sixteen Super Frelons, while there are forty Alouette IIIs operated on liaison and AOP duties, and for training. The South African Navy operates ten Westland Wasps from four frigates, but these helicopters may be relegated to the liaison role or abandoned altogether in the near future as the warships are overdue for replacement and once retired are likely, due to the embargo, to be replaced by South African-built coastal defence patrol boats of small size, rather than by vessels of similar size to those currently in service.

There are a number of sizeable naval and military air arms in Latin America, although few maintain a substantial element of modern equipment, partly due to economic problems and partly because of attempts by the United States to limit military supplies and encourage expenditure on social welfare schemes rather than defence. While there is no direct threat to the nations of the region, sheltered as they are by the United States, there is some involvement by Cuba with support for Communist-inspired insurgents in certain countries, while elsewhere, a difficult terrain and large areas ensure that there is a continuing need for military aviation, and for naval aviation to patrol long coastlines and massive exclusive economic zones stretching up to 200 miles from the shoreline. To some extent, the helicopter takes second place to nationally built fixed-wing counter-insurgency aircraft, which save foreign exchange and create local employment opportunities, and are cheaper to operate.

One of the strongest Latin American powers is Argentina, whose air force, the Fuerza Aerea Argentina, operates fourteen Hughes 500M helicopters on counter-insurgency duties, with six gunship Bell UH–1H Iroquois, and in the difficult Antarctic region, there are three Sikorsky S–61s and three Boeing-Vertol CH–47C Chinooks on transport duties, complementing a force of Bell UH–1Ds, 212s and Sikorsky UH–19s operated elsewhere in the country on liaison, utility and transport duties, while four Bell 47Gs are employed as trainers. The Argentinian Navy operates eighteen Westland Lynx anti-submarine helicopters from two British-designed destroyers and from four frigates, and some of these aircraft join the five anti-submarine Sikorsky S–61D Sea Kings aboard the Navy's single British-built light fleet aircraft-carrier, while the Navy also operates a few Pumas and Alouette IIIs on transport and liaison duties. The Argentinian Army maintains its own air arm, with up to twenty Bell UH–1Hs, two Boeing-Vertol CH–47C Chinooks and twelve Aerospatiale Pumas on transport duties, as well as some Agusta A.109s, Bell JetRangers and FH–1100s on liaison and training duties. A number of these aircraft will have been lost during hostilities with Britain over the Falkland Islands.

The largest armed forces in Latin America are those of Brazil, with an Air Force which includes a counter-insurgency element with forty Bell UH–1Hs and a small number of 206s and Hughes OH–6As, as well as six Aerospatiale Pumas on SAR duties and six Sikorsky SH–1Ds. The Forca Aeronavale operates six Sikorsky S–61D Sea Kings, and nine Westland Lynx anti-submarine helicopters from the six Niteroi-class frigates, as well as some older Westland Wasps and Whirlwinds, and eighteen Bell 206s, some of which are used for training while the remainder are employed on liaison and communications duties.

Other countries in the region fare less well. Chile, for example, has had to face arms embargoes in the past from a number of countries, including Britain, and today operates in her Air Force an assortment of helicopters, including small numbers of Pumas and Lamas, and Bell UH–1Hs, while the Navy operates Alouette IIIs, Bell JetRangers and just two UH–1Ds and fourteen Bell 47s, on SAR, liaison and training duties, including operations from two Leander-class frigates. The Chilean Army operates nine Pumas, three Bell UH–1H Iroquois, and some JetRangers and Lamas on liaison and utility duties. At the opposite end of South America, Columbia operates a more modern force, with the Air Force maintaining twenty-seven SAR Lamas, sixteen training Bell 47Gs, and on behalf of the Army, an AOP element which includes twelve Hughes OH–6As and ten 500Cs, plus Kaman HH–43B Huskies, Sikorsky S–55s, Bell UH–1Bs and Hiller H–23s.

Oddly, two of the better equipped nations in South America are two neighbours, Peru and Ecuador, both of which maintain reasonably well-equipped and sizeable armed forces, tending to operate some of the most modern weaponry in Latin America. No doubt a degree of mutual antagonism lies behind this mini arms race, and indeed there

222

have been border clashes between the two countries, including a minor war as recently as January 1981. However, the Ecuadorean Air Force maintains a fairly small helicopter element, with just two Pumas on VIP transport duties, while there is also a very small number of Lamas, Alouette IIIs and Bell 212s. Peru operates six Agusta-Bell 212 anti-submarine helicopters in a small naval air arm, operating from Italian-built Lupo-class frigates, as well as four Agusta-built SH–3D Sea Kings on ASW work, and some Bell JetRangers, 47Gs and UH–1Ds and two Alouette IIIs. However, the main helicopter operator is the Peruvian Army, with some forty-two Mil Mi–8s for transport duties, and a small number of Bell 47Gs and Aerospatiale Alouette IIIs.

Australia's armed forces have suffered from alternating periods of strong official support with re-equipment and periods of neglect, often based on the argument that Australia faces no immediate threat to her national security. However, Australia is allied with both the United Kingdom and the United States in the South East Asia Treaty Organization, although this lacks the cohesive command structure and full-time secretariat of NATO, and with both these countries in separate defence treaties as well as with Singapore and the Federation of Malaysia in certain bilateral agreements. New Zealand is also a party to many of these agreements, and in spite of the argument that there are few obvious threats to the region, there has been a growing Soviet maritime presence in the Pacific and Indian Oceans, and the threat on the edges of the region, in South-East Asia, is too obvious and well-established, and indeed even successful, to be ignored.

The Royal Australian Air Force is primarily a fixed-wing aircraft operator, but it does maintain a squadron of twelve Boeing-Vertol CH–47C Chinooks and some forty Bell UH–1Hs on transport and utility operations, as well as some SAR, mainly on behalf of the Australian Army. The Australian Army itself has some fifty Australian-built Bell 206s and a number of fixed-wing aircraft for close support, AOP and liaison duties. The Royal Australian Navy operates Westland Sea Kings on anti-submarine duties from its sole light fleet carrier, HMAS *Melbourne*, as well as five Bell UH–1Bs, some 206s and a few remaining Westland Wessex anti-submarine helicopters, some of which are in reserve. It is possible that one of the new British Invincible-class through deck cruisers, may be ordered for the Royal Australian Navy in the near future, and this ship will operate the Sea Kings. Less certain is which helicopter will be chosen for the four new frigates of the Oliver Hazard Perry-class which the Royal Australian Navy is receiving, and which will be its first helicopter-carrying escort vessels; probably these ships will either operate a single Sea King each,

A Royal Malaysian Air Force S–61 above the jungle.

or the decision may be taken to obtain Sikorsky Sea Hawk helicopters to ensure greater commonality with the United States Navy, on whose expertise and equipment the RAN now depends heavily.

In New Zealand, with a small population and little more than a token attempt at defence, the Army does not operate aircraft, but instead is the main user of fourteen Royal New Zealand Air Force Bell UH–1D Iroquois transport and utility helicopters, which are backed by some Bell 47Gs devoted to the training role. The small Royal New Zealand Navy operates four frigates, one of which is devoted to training duties, and the other two, both British-built Leanders, HMNZS *Canterbury* and *Waikato*, operate the RNZN's two Westland Wasp anti-submarine helicopters. A new frigate may be ordered, but it is more probable that one of the older ships will be modernized, with conversion to gas-turbine propulsion.

Clearly, today few air forces or air arms can manage without the helicopter, but it is also true to say that the helicopter still remains a relatively expensive item of equipment, which puts it outside the budgets of the poorer nations, even though these could often use its unique capabilities on disaster relief work. The role of the helicopter has grown throughout its history as its capabilities have increased, but it is also interesting to note that many older machines linger on in service in the poorer nations, demonstrating possibly rather greater reliability than might have been considered possible in this complex machine at the time of manufacture.

10
INTO MUFTI

Unlike most aircraft types, there has always been a close relationship between the helicopters used in civilian or commercial operations and their military or naval counterparts. While this type of relationship has also held true for most fixed-wing transport aircraft, it takes no great understanding of the subject to appreciate that there is today a wide gulf between combat fighter or interceptor and bomber aircraft, and almost any type of civilian aircraft imaginable. Of course, the gap has not always existed even for fixed-wing aircraft and, at the outbreak of World War II, there were many aircraft types entering service as bombers, especially with the Luftwaffe, which had been developed as civil airliners, albeit ones with cramped and inconvenient fuselage cross-sections. Even at the end of World War II, the aerodynamic surfaces of such aircraft as the Avro Lancaster bomber and Armstrong-Whitworth Wellington bomber were used for such hasty civil airliner designs as the Avro York and the Vickers Viking. A close link still remains between civil airliners and maritime-reconnaissance aircraft. This said, there can be no such comparison between interceptors such as the McDonnell Douglas F–15 and airliners such as the Boeing 747, the famous jumbo jet.

There are several reasons for the continued close connection between helicopters intended for combat use and those with a more peaceful role. The most obvious must remain that in spite of the many uses to which the helicopter can be put, transport and utility duties still account for most of the helicopters sold whether to military or civil customers. The other reason must be that the speed of the helicopter is still sufficiently low for military and civilian applications to be contained within the same general level of performance, and that the specialization for most helicopters is accounted for not by power-plant or aerodynamic changes, but by changes in equipment and in particular electronic equipment. From this, of course, it is immediately obvious that the rift which has occurred with the development of the high performance fixed-wing combat aircraft is now also occurring with the rotary-wing machine, with the advent of such specialized types as the attack helicopter, designed to provide the smallest and most difficult to detect fuselage cross-section, the maximum external warload and the best speed.

Of course, the helicopter still possesses considerable disadvantages in terms of first cost, maintenance costs and fuel consumption compared to fixed-wing aircraft, and for this reason, its spread into commercial operation has been slower than would otherwise have been the case. The customer has to pay a premium for the unique capabilities of the helicopter. One further reason for the connection between civilian and military helicopters arises from this, since sometimes the civil market for a particular design has been too small to justify the development of a separate helicopter, and so helicopters have been developed on the back of a military requirement, with the civil version extending the production run.

There are limits to what the civil market can accept, however, and one non-starter as far as civilian applications were concerned was the proposed civil version of the Sikorsky S–56, simply because few, if any, civil operators could have justified the cost of such a large and expensive twin piston-engined machine, even allowing for the lifting capabilities. Indeed, few civil helicopters have matched the size of the largest military machines, other than in the Soviet Union, where the distinction between military and civil aviation is blurred, with Aeroflot, the state airline, being managed on military lines, and where the poor surface communications and hostile environment of much of the country encourages use of the helicopter. On the other hand, some basically military or naval designs have been worth redesigning to create a civil version, with a larger cabin to make the most of the available power. One of the best examples of this approach has been the creation of the Westland WG.30 from the power-plant and rotorhead of the Lynx helicopter, while up-rated engines are also available to boost the performance of this helicopter still further, to the extent that it may even, if the customer demand arises, be able to match the performance of the Sikorsky S–61, but in a rather more modern package with the new Rolls-Royce Gem turbo-shaft.

Nor has the traffic been all one way. The Bell 206 JetRanger, the Aerospatiale Dauphin and the Sikorsky S–76 Spirit are all examples of essentially commercial designs which have found their way into the military market, and even the WG.30 is likely to find many military customers in the years ahead.

The original production helicopter, the Sikorsky R–4, would have been at home with a civilian operator, had such existed and had a role been apparent, but for the wartime condition into which this small machine was born. The presence of a civil market and the potential which it offered was not ignored, and indeed it was as obvious to the pioneers as was the military and naval market, and this was especially true of the helicopter's role as an air taxi at a time when such heavy work as flying-crane operations were out of the question.

The Sikorsky S–51, the R–5 of the United States Army and the WS–51 Dragonfly in its British form, made up for the time wasted during the war years. Civilian operators and postal administrations were soon keen to try the new aircraft type. One of the earliest examples occurred when the then new British European Airways, which was eventually merged with the British Overseas Airways Corporation to form British Airways, commenced a series of trials using Dragonflies to carry mail between a number of destinations in the west of England. The airmail experiments were run for a period during 1947 and, three years later, BEA inaugurated the world's first regular helicopter passenger service, flying for a short period between Liverpool and Cardiff. Other Dragonfly customers included the Belgian airline, Société Anonyme pour l'Exploitation de Navigation Aérienne, or Sabena, while the aircraft in both its Sikorsky and Westland forms began to be used to check electricity transmission cables. The development of the WS–51, the Westland Widgeon, also sold to civilian customers.

A small number of these early helicopters found their way into police use in various parts of the world, although lacking modern traffic problems, the police forces concerned tended to be those with more of a para-military role, including, for example, the Royal Hong Kong Police, which purchased two Widgeons, used primarily for border patrol and detection of illegal immigrants and smugglers.

The distinction of being the first commercial production helicopter to receive a civil certificate of airworthiness, awarded by the United States Civil Aeronautics Administration on 8th March 1946, belongs to another successful helicopter type, the Bell 47. The Bell 47 rapidly became *the* helicopter during the late 1940s and throughout the 1950s and into the early 1960s for those needing a small 'air taxi' for commuting between offices or factories, or from these to the nearest civil airport. It also proved invaluable to companies involved in either forestry, oil exploration and production, or in extraction of other raw materials, reflecting the often difficult surface terrain and poor transport infrastructure in many parts of the world. Bell 47s were also the pioneering helicopters for aerial crop-spraying, removing the need to provide a landing-strip when planning such operations, while one company, Continental Copters, developed a special single-seat agricultural helicopter based on the Bell 47 airframe, power-plant and mechanical systems, known as 'El Tomcat'. Specifically designed for the civil executive transport market, the Bell 47J Ranger, which

227

entered production in 1954 and offered a four-seat light aircraft-type cabin in place of the familiar goldfish bowl, used an up-rated 220-hp Lycoming VO–435 engine, and remained in production until 1966, when it was replaced by the highly successful JetRanger. The standard model of the 47 remained in production until 1974, a total production life of twenty-eight years.

Agusta in Italy manufactured the 47 series under licence, including the 47 Ranger, and even developed some more powerful versions of this machine, ideally for use in Alpine regions where higher performance is at least useful and often essential.

In spite of problems with vibration and with the shadow of the rotor blades, the helicopter also became known at an early stage as an ideal platform for filming, being able to remain in the hover, move backwards or sideways, or even slowly forwards, as necessary, and generally being able to use of all its special characteristics to establish an unrivalled superiority over the fixed-wing aeroplane except, of course, for air-to-air photography of fast-moving aircraft!

The success of these two early designs, the S–51 and the Bell 47, obviously convinced the manufacturers that useful additional business was available providing that the civil market was borne in mind. It was, as we have already noted, not difficult to adapt basic military designs for the commercial operator, many of whom had a rugged utility requirement rather than a need for a plush and sophisticated transport, and even the airlines appreciated that they were dealing with an experimental machine for use on new routes, hitherto unserved. Passengers were too taken with the novelty of vertical take-off seriously to consider comfort, and the convenience of course permitted all manner of shortcomings in other areas. It should also be remembered that few passengers had, even during the mid-1950s, sampled the delights of jet or turbo-prop airliner travel, offering smoothness and speed, and those who had were mainly in Britain and Europe, since it took far longer for the turbo-prop to find acceptance in the American air transport scene. Most American airlines offered piston-engined air travel, with only one using the revolutionary British Vickers Viscount airliner, although this aircraft was used by most of the major European airlines, including BEA, Air France, Alitalia and KLM.

The successor to the S–51, the Sikorsky S–55, or Westland Whirlwind in its anglicized form, showed obvious advantages in the way in which it moved troops around the Korean battlefields, and American civil certification was granted on 25th March 1952. An early commercial, or at least non-military honour for the Westland

Whirlwind was that it became the first helicopter to be accepted as suitable for the Queen's Flight, the RAF-operated service which, in spite of its name, also carries members of the British Government in addition to the Royal Family. The VIP versions generally provided between four and seven seats, against the eight or ten seats of normal transport versions. There were other distinctions too, including that of being the first helicopter to operate international scheduled services when, in 1953, the Belgian airline, Sabena, inaugurated these services using Sikorsky S–55s, providing fast transport between central points in the densely built-up area of Belgium, the Netherlands and northern France. BEA used the Whirlwind to provide experimental city centre to London Airport services, operating these flights under the aegis of the BEA Experimental Helicopter Unit, which was formed in 1947 and remained until the formation of BEA Helicopters in 1964. Difficulties over helicopter noise limited the scope and the duration of these experiments.

The value of the helicopter was fast growing, and specialized helicopter operators began to appear, such as Autair and Fisons Airwork, and with the advent of larger helicopters, the flying-crane role, ferrying supplies over difficult terrain, or perhaps neatly placing equipment or an element in a construction project into place, began to come to the fore. A rather more successful and consistent operator of city centre to airport services arose with New York Airways, a specialized helicopter operator formed in 1949 and which pioneered this type of operation, although no less persistent have been the environmental and safety objections to helicopters flying over densely built-up areas of population.

Not all of the successful commercial helicopters were American or licence-built American designs. The success of the Sud, later Aerospatiale, Alouette II and III series with military users was matched by a corresponding success with commercial customers, with these helicopters fitting neatly between the small Bell 47 and its derivatives and the larger, and more expensive, helicopters of S–55 size. Indeed, in 1963, the distinction of being the first turbine-powered helicopter in commercial service in the United States fell to an Alouette, with the Alouette II being assembled in the United States by the Republic Aircraft Corporation, now part of Fairchild. The hybrid Aerospatiale Lama, with the airframe of the Alouette II and the mechanical components of the Alouette III, has been sold mainly to commercial customers.

In the Soviet Union and the Soviet satellite nations, both the Mil and the Kamov family of helicopters have been pressed into commercial service with the state airlines. The Kamov series, starting with the Kamov Ka–15M, first flown in 1952 but possibly not available for commercial use until some three or four years later, was used on air taxi, forestry and fisheries protection patrols, and for agricultural crop-spraying, as indeed was the Ka–18, which followed a few years later. A successor to these early Kamov machines, the Ka–25 included a flying-crane version capable of lifting up to $2\frac{1}{2}$-ton loads and intended primarily for civil duties, while the current production machine, the Ka–26 is also available for the full range of agricultural and patrol duties and, with special electronics, for geological survey duties. The large Mil flying-crane helicopters are also much more in evidence in the Soviet Union than their Western counterparts are in the United States, overcoming some of the drawbacks of the poorer Soviet infrastructure, and helping with massive civil-engineering projects on a far more regular basis than would be the case in the West. The broader expanses of the Soviet Union, the lack of public opposition, or the opportunity to give voice to that opposition, to noise, the lower infrastructure base and the penchant for grandiose projects at the expense of other matters, must all contribute to this greater utilization of the helicopter. However, it may be no small matter either that Aeroflot and its satellite sister airlines are run more on military than commercial lines, and that the massive civil helicopter fleet is in effect an organized reserve for the Soviet armed forces.

An agricultural version of the Kamov Ka–26, for crop spraying.

A geological survey Kamov Ka–26.

A Mil Mi–6 demonstrates its capabilities for fighting forest fires.

A Mil Mi–2 in civil guise.

Returning to the United States and to Sikorsky, still greater penetration of the civil market, or perhaps one might even suggest, development of that market, occurred with the next helicopter, the S–58 series, which offered a significant improvement in performance against the S–55. Civil certification of the commercial Sikorsky S–58B and S–58D series came in 1956, and a substantial number of the 1,800 or so S–58s built by Sikorsky, and of the Wessex development built by Westland, were produced for civil customers. Although capable of lifting up to sixteen troops, the commercial passenger or airline versions of the S–58 tended to be mainly ten- or twelve-seat machines, and the American-built aircraft for such customers as Sabena, which took eight, were somewhat unusual in having two sets of doors, allowing entry into the passenger cabin in much the same pattern as with the older European suburban and branch line railway carriages. Other customers included New York Airways, with three, Chicago Helicopter Airways, with eight, and the Canadian operator, Okanagan Helicopters, while Bristow Helicopters, at the time part of British United Airways, operated a total of eighteen of the Westland Wessex in Britain and on detachment elsewhere, including Africa and the Far East.

The S–58 and the Wessex became available at the start of oil exploration and production offshore, initially in the Gulf of Mexico and then, rather later, in the North Sea. Many of Bristow's helicopters also helped the support of exploration efforts for oil and minerals in the Far East, and Okanagan Helicopters performed a similar function in the frozen wastes of northern Canada. For a spell, Bristow Helicopters allocated a Wessex to a Department of Trade contract, providing an air-sea rescue service off the Kent coast following the withdrawal of RAF SAR helicopters from Manston as a result of defence cuts, and which left the busy shipping lanes at the northern end of the English Channel and at the Thames estuary without helicopter cover. After several years of satisfactory service, the Bristow machine was replaced by the restoration of the RAF service. Provision of SAR services by civilian operators under contract is not the only blurring of the distinction between civil and military helicopter activity, since operating to and from oil exploration or production platforms has something of the element of ship-board operation about it, landing on the small helicopter platform which is a feature of almost every oil or gas rig today, and which, while they may not roll and pitch as much as a small escort vessel, do not always provide a stable platform in very bad weather since the most modern types are of semi-submersible construction. Most oil-rig platforms provide no protection at all from the worst that the elements can offer and, on several occasions, the offshore support helicopters have suddenly had to turn to rescue work, evacuating personnel from a rig which was endangered.

The closure of the Suez Canal in 1967 accelerated the trend towards larger oil tankers, the so-called very large crude carriers or VLCCs, too big to squeeze through the canal and carrying Europe's oil from the Middle East via the Cape of Good Hope. Such large ships could not call at ports *en route*, or *en passage*, to collect supplies or change crews on their long voyages, and so the helicopter again found a new role, providing support for these vessels, ferrying out supplies or crew members, or owners' agents or representatives. In the more confined waters closer to Europe, such large vessels often required the assistance of a pilot earlier than cargo ships, which were often a twentieth of the size of the tankers, partly because of the danger of beaching on shallows and partly because of the wider turning circle and longer stopping distance of the super tankers. Here too the helicopter came to the fore, transporting pilots to the tankers, or taking pilots off on the outward sailing, replacing the old pilot cutter and increasing productivity and the actual duty period spent aboard the ship for the pilot. It is not always necessary for helicopters flying

233

out to tankers to be very large, and often the smaller executive types are used, in contrast to the offshore oil support helicopters, which are usually medium- or heavy-lift types, carrying small airliner passenger loads and operating over far longer distances than the tanker support helicopters.

Sikorsky's position as the major supplier of helicopters for the heavier utility duties and for offshore support was consolidated during the 1960s and 1970s following the introduction of the S–61 series, with the land-borne S–61L receiving civil certification from the FAA in 1960, and the amphibious S–61N, normally used for offshore support, being certificated in August 1962. Both types have the amphibious boat-type hull form, as have the naval versions, but the amphibious S–61N has the stabilizing floats into which the undercarriage can retract, and both versions have a completely different rear cabin shape to the military and naval versions, allowing rather more passenger space for up to twenty-eight passengers in addition to the crew of three. Entering production and service at the time when offshore oil exploration was developing fast, the S–61 series has been produced in very large numbers for civil customers, including British Airways Helicopters, the successor to BEA Helicopters, KLM-Noordzee Helicopters, Helicopter Service of Norway, Bristow Helicopters and many others. However, it was an S–61L which first entered service, commencing operations with Los Angeles Airways on 1st March 1962, and later also flying for Carson Helicopters and New York Airways, while most subsequent customers seem to have favoured the greater flexibility bestowed by the wider capability of the S–61N.

Even with the capability of the S–61N, the commercial helicopter in regular passenger service has had mixed fortunes, with one of the very few successful passenger services being that between Penzance in the south-west of England and the Isles of Scilly, off the Cornish coast, operated in 1964 by the former BEA Helicopters. Even before the S–61 series arrived, Sabena scrapped its network of helicopter services, suffering from competition from private cars and improved rail services as well as continuing losses due to the high cost of helicopter operation. Pakistan International Airways introduced a network of service in East Pakistan, now Bangladesh, providing vastly improved air-transport facilities during the 1960s in an under-developed area with poor road and rail communications. However, services either became too costly for continued operation to be justified or, on other routes, became sufficiently popular to justify a basic landing-strip and the lower costs, and higher productivity with higher block speeds, of the conventional airliner, and the change to airliners

234

was made in 1966, after two accidents had reduced the helicopter fleet from three aircraft to one. To some extent, this has been the dilemma of the helicopter in commercial operation, with success creating the justification for an investment in airport facilities. The aircraft manufacturers have also proved themselves more than equal to the challenge of producing a steady succession of ever more capable airliners offering short take-off and landing capability at lower first and lower operating costs than the helicopter. One of the most successful manufacturers in this field, if not the most successful, has been de Havilland Aircraft of Canada, at one time a subsidiary of the British Hawker Siddeley Group, which had on the formation in 1961 of Hawker Siddeley Aviation absorbed Avro, one of Cierva's licensees in Britain, but other manufacturers have since also competed with DHC. Even the Penzance and Isles of Scilly service has had competition, from elderly de Havilland Rapide biplane airliners, Britten-Norman Islanders, and today, from de Havilland Canada Twin Otter aircraft.

Boeing-Vertol's predecessors devoted most of their energies to the development of naval and military helicopters until the advent of the Vertol 44, a twin-rotor piston-engined helicopter which received FAA certification in 1957, and which was subsequently sold in small numbers to New York Airways, which leased two to Sabena in 1958, and to a few other smaller operators, including some oil companies, while the Royal Canadian Air Force purchased five of the civil version, and these were operated on its behalf by Spartan Air Lines in Canada. The small fifteen-seat helicopter did, however, provide a useful toe-hold in the airliner market for Vertol, and to some extent its success was compromised by the development and pending production of the turbine-powered Vertol 107, which tended to blight sales of the earlier aircraft. Certification of this twenty-five passenger helicopter was in January 1962, designed with the needs of New York Airways in mind, with that operator flying three helicopters leased to it by Pan American Airways, followed later by three Kawasaki-built models: after 1965 all civil versions were in fact built by Kawasaki, which sold a number of 107s, known as the KV–107/II to operators in the Far East.

After the acquisition by Boeing of Vertol, the attitude towards civil sales became rather more positive, as one might expect of the world's largest commercial airliner manufacturer. Boeing-Vertol is today actively pushing sales of its CH–47D Chinook to civil customers, and has already gained such important customers as British Airways Helicopters. The Chinook offers up to forty airliner seats, and with its speed and range as well as this capacity, is possibly the first large helicopter actually to challenge the fixed-wing airliner effectively on shorter journeys. The challenge arises from the fact that before its introduction, oil-rig workers would be taken off by an S–61N and flown to an airfield such as Sumburgh, in the Shetlands, to be transferred to a British Aerospace 748 twin turboprop airliner for the flight south to Aberdeen, but the Chinook can fly many such charter contracts on a direct oil rig to Aberdeen basis, cutting out the transfer and overall reducing costs, often with one Chinook able to handle the work of two S–61Ns or two Aerospatiale Pumas.

The only non-American helicopter, or non-licensed American helicopter design, to have entered the larger end of the civil market with any real success so far has been the Aerospatiale Puma, and this too is a civil derivative of a military machine, designed by Aerospatiale and built jointly by the French Company and by Westland Helicopters in Britain, which produces a large number of airframe parts. First flown on 26th September 1969, the civil Puma uses less powerful engines than the military version, there are two 1,290-shp Turbomeca Turmo IVA turboshafts against the 1,328-shp of the Turmo IIIs in the military variants, and is designated the SA.330F, with accommodation for about sixteen passengers. Amongst the civil customers for the SA.330F can be included Bristow Helicopters, the large British helicopter operator which is in fact the world's largest commercial helicopter operator, although its seventy or so aircraft appear insignificant against the 9,000 or so of the US Army, the largest military helicopter operator!

Aerospatiale has, however, also maintained its stake in the lower end of the civil market, with civil developments of the SA.341 Gazelle light helicopter for air taxi and other uses, with five seats. Most of the nine hundred or so Gazelles produced so far have been sold to military customers, but there have been substantial civil sales nevertheless, while the civil market itself has developed sufficiently for Aerospatiale to develop and market successfully such specifically civil small helicopters as the Dauphin and Dauphin II, and in a reverse of the normal pattern, these can be modified to suit military customers; the main customer in this category so far being the United States Coast Guard Service.

A challenge to the Puma, and its up-rated development, the Super Puma has come recently from the Westland WG.30, which uses the basic Lynx engines, gearbox and rotor systems, but with a much larger cabin and which accommodates up to sixteen passengers. There is the possibility of an up-rated version being able to compete effectively on certain aspects of performance with the S–61 series, which is now out of production.

The United States Army's LOH or Light Observation Helicopter competition, held during 1962 and 1963, was largely responsible for providing the civil market with two of its most successful light helicopters to date, with the Bell 206 JetRanger, at one time considered a losing finalist, although within a few years this machine was in US Army service in very large numbers, and the winning Hughes 500. Another loser, the Fairchild-Hiller 1100 also enjoyed considerable success on the civil market, with its achievement diminished only by the unprecedented success of the 206 and the 500. Some 6,000 JetRangers have been built by Bell and its licensees, including Agusta in Italy, and mainly for the civil market, these helicopters have now been joined by a stretched version with up-rated engines, the Bell 206L LongRanger, offering seven seats instead of the five of the original JetRanger and of the Hughes 500 and Fairchild-Hiller 1100. There can be no doubt that the helicopters of this generation have stimulated demand for the helicopter from business users, offering speeds close to those of small business aircraft enhanced by the helicopter's vertical take-off capability. Costs are higher than for comparable business aircraft, but for many users, the decision to buy a helicopter has been made easier by the tendency to opt for a helicopter with all of its advantages rather than for a larger or faster business aircraft. The use of turboprop propulsion on a growing number of light aircraft has increased their costs, while piston-engined aircraft have suffered severely from the shortage of gasoline supplies in many parts of the world, and its high cost, about £1.00 ($2.00) more than a gallon of petrol in the British Isles at the time of writing; this also makes the decision to buy a turbine-powered helicopter easier. For many users, it is a choice of a helicopter with shortened journey times over short and medium distances because of its ability to operate from factory car-parks and other such suitable places, and using it as an air taxi to fly to airports to join scheduled flights, or of a larger and more sophisticated fixed-wing business aircraft, capable of flying the longer international distances, but lacking the helicopter's flexibility. There has also been an acceleration in what may be considered as the natural growth in the civil helicopter market.

Between the helicopters of the 206 or 500 size and the Puma and S–61 types, there lie helicopters of which the best known by far is the Bell 204 and 205 series, the Iroquois. While best known as military helicopters, the 204 and 205 series have also sold to civil customers, and the introduction of the twin turboshaft development, the 212 and the AB.212, has in fact developed the market still further, with the

Bell 212 being relatively far more popular with civil customers than with the military, although this may be due largely to the saturation of the military market with 204 and 205 series machines, especially after the end of the Vietnam War left the American armed forces in particular with a large volume of equipment at a time of contraction as conscription was phased out in favour of professional armed forces. The main military users remain the US armed forces, and those of Canada, for the 212, while the Spanish and Italian navies also use this type, but elsewhere, an even more powerful version, the 214, has entered military service, although this type has seen little civilian use.

Bell too is also considering the civil market in an increasingly specialized manner, with needs of its own, and is capitalizing on its dominance of the smaller end of the helicopter market with the new Bell 222, intended for civilian customers, and in common with the latest Aerospatiale designs, treating military customers for this aircraft as a variation on its original concept.

Elsewhere, some helicopters, such as the MBB BO.105, designed and built in West Germany, have been developed with both the military and civil markets in mind, while the MBB and Kawasaki Mk. 107 is primarily a civil helicopter. Further up the scale, the need to consider the civil or commercial market as being worth while in its own right is finding increasing acceptance, and in developing the next generation of anti-submarine helicopter, the EH–101, the two partners in this European project, Westland in Britain and Agusta in Italy, are bearing the commercial customer in mind from the start, so that this will not only be a replacement for both Westland and Agusta-built Sea Kings, but also potentially for the civil S–61Ns and Pumas as well. Agusta's new A.109 Hirundo, Swallow, is a civil helicopter which is also finding favour with a number of armed forces throughout the world.

The rising cost of fuel and the continuing objections from increasingly vocal environmentalists over the noise of helicopters operating from city-centre landing sites have together effectively cancelled the more ambitious plans for helicopter development, or, because the helicopters exist, perhaps one should say for the further development of helicopter passenger services. Plans mooted at various times for passenger services from the centre of London to the centres of Paris, Brussels and Amsterdam, using either Sikorsky S–65s or Boeing-Vertol CH–47D Chinooks, charging passengers fares equivalent to the existing airport-to-airport first-class rates, and offering a far faster city-centre to city-centre journey time, have been stillborn. In many ways, this must be counted as a loss, since such services would have a good chance of viability, with the presence of the English Channel creating an obstacle in the development of surface transport which even a Channel railway tunnel or bridge would not overcome completely in terms of journey time. The existence of the English Channel in itself has probably been responsible for much of the strength of Britain's airlines today, which are generally far larger than their European counterparts, which have faced strong road and rail competition during their development. Indeed, at the time of Britain's joining the European Community, the total British air transport effort in terms of aircraft and aircraft ton-miles, was almost equivalent to half the total effort for the original six members, which included such large countries as Germany, France and Italy.

It might be that services between Paris and Brussels, Paris and Amsterdam and Amsterdam and Brussels would stand less chance of success because of the presence of such fast and well-established competition; most frequently the door-to-door convenience of a fast motor car.

The ability to operate such major international services is not recent, for had the original Fairey Rotodyne entered production, that aircraft would still be able to provide such services more quickly and with larger passenger loads than any helicopter in service in the West today. During the early 1970s, considerable attention was given to the possible development of vertical take-off transport aircraft, and it seemed at one stage that the enthusiasm of European manufacturers for this concept, coupled with Britain's success in building the first operational vertical take-off fighter aircraft, meant that such development would allow the Europeans to eliminate the lead which the Americans and Russians have established in helicopter production. However, the Arab oil embargo and then rapidly rising oil prices from 1973 onwards, coupled with a downturn in many Western economies, made this development seem costly and even extravagant, and certainly unnecessary, while attention turned to developing quieter and more fuel-efficient aircraft. The choices open to the manufacturers included tilt-wing turbo-shaft-powered aircraft, and various types of turbo-fan-powered aircraft with banks of eight or ten lift jets, but while some of these might have been faster than the Rotodyne concept, none offered the relative simplicity of that design, although several might have offered higher speeds.

If vertical take-off ever does reach the commercial market in a form other than that of the helicopter, it is almost certain that it will again be as an offspring of a military or naval machine, built to support either armies in the field or warships able to operate only vertical take-off

239

aircraft. The cost of development is now too high and the benefits too little known for any civil design to stand a chance on its own. The technology exists, but needs to be pulled together and developed, and for that the money is no longer available.

Either the compound helicopter or the advancing blade concept, of which we will see more in the next chapter, might offer the next step forward for the helicopter, but both have to be proved in regular military use even now, and since neither has really developed beyond the experimental stage, this means that they are far off, while so far such development as has been made has been aimed mainly at future attack helicopter designs, and not the larger transport and utility helicopters. It can still be argued that since the compound helicopter combines rotary-wing propulsion and conventional propulsion, then both the Rotodyne, and the Russian Mil Mi–12 twin rotor machine of 1971, can be regarded as forms of compound helicopter, but neither entered production.

11
THE FALKLANDS

Although the helicopter has been the aircraft type used in the greatest numbers in many of the conflicts which have beset the world since the end of World War II, no one instance can have shown the helicopter operating in so many different roles as the recent hostilities in the South Atlantic between British and Argentinian forces in and around the Falkland Islands and South Georgia. The conflict was notable not only for the variety of tasks handled by the helicopter, but also for the identification of other, newer, roles to add to what is already an impressive list. By coincidence, the hostilities in the South Atlantic during April, May and June 1982, overlapped with a fresh outbreak of hostilities in the Middle East, as Israeli forces invaded the Lebanon in pursuit of terrorists belonging to the Palestinian Liberation Organization, and found themselves also in conflict with Syrian forces.

Although British territory for about 150 years and settled continuously by settlers of British descent for most of that time, Argentina had for long claimed sovereignty over the Falkland Islands, based on a claim of prior settlement, which was subject in itself to dispute because the period of the Argentinian claim was one of considerable movement in the allocation of colonial rights, and the Argentinians who had arrived on the islands, which had been previously visited by British and French explorers, lacked official status. Other grounds for the Argentinian claim included the fact that the islands were part of the South American continental shelf, even though this could hardly be regarded as a major factor in the sovereignty of islands some 400 miles off the coast! For some years, the British had attempted to negotiate some form of solution with the Argentine Republic, but to no avail. In spite of the islands being some 8,000 miles from Britain, a small British community had long established itself on the islands, although since the decline of the whaling industry, South Georgia had been populated only by scientists and naturalists; the Argentine also laid claim to South Georgia, some 1,500 miles from mainland Argentina, which was a separate British dependency administered from, but not part of, the Falklands.

While negotiations continued over the future of the islands, in which some form of practical compromise was sought by successive British governments, in spite of firm resistance by 1,700 islanders to

Argentine attempts at control of the islands, the Argentine planned an invasion. There is some doubt over the timing of the decision, which many believe may have been encouraged by the British Government's decision to withdraw the Royal Navy's Antarctic patrol vessel, HMS *Endurance*, and was preceded by the landing of a party of scrap-metal merchants on South Georgia without British permission or the completion of any kind of immigration formalities.

Argentina's military and naval helicopters played relatively little part in the initial invasion of the Falkland Islands on the morning of Friday, 2nd April 1982. The bulk of the Argentine force arrived by landing-craft, battling against fierce resistance by a small force of some seventy Royal Marines who were eventually forced to surrender in the face of several thousand Argentine troops. There appear to have been no landings of troops from the Argentine aircraft-carrier, *Veinticinco de Mayo* (25th May), formerly the Royal Netherlands Navy *Karel Doorman* and before that the Royal Navy's HMS *Venerable*, although this ship remained off the coast of East Falkland ready to provide air and anti-submarine cover, but it would seem that the military governor of the Falklands arrived at Port Stanley by helicopter from the ship. Reinforcements for the initial invasion force arrived by sea and by air, using Lockheed C–130 Hercules transport aircraft augmented by requisitioned civilian Fokker F–27 Friendship airliners.

The following day, Argentine marines invaded the British dependency of South Georgia, more than a thousand miles east and south of the Falklands. An Aerospatiale Puma and two Alouette III helicopters flown from the ice-patrol vessel, *Bahia Paraiso*, 9,600 tons, to land troops and suppress defensive fire from the island's garrison – just twenty-two Royal Marines – ran into difficulties almost immediately. The Royal Marines just missed shooting down the Puma using a 66-mm anti-tank gun, which missed the helicopter by inches, but then machine-gunned the Puma until it retreated across a bay and crash-landed, and then a few minutes later one of the Alouettes was shot down. Shortly afterwards, the Marines were forced to surrender, astonishing the invaders who found it hard to accept that they had nearly been defeated by a token defending force, which had also used its anti-tank weapons to cause severe damage to an Argentinian corvette! Such an unorthodox approach to the use of anti-tank weapons could cause some reconsideration of the role of the helicopter on NATO's central front!

The role of the helicopter in the operations around South Georgia did not end with the Argentine invasion, since the helicopter played an important role in the retaking of the island.

Following the invasion of the Falkland Islands, the British Government ordered a Royal Navy task force to sail to the area, taking a sizeable detachment of Royal Marines and British Army units, while diplomatic pressure was applied to force an Argentine withdrawal – although diplomacy failed completely. The task force, which was prepared for sea within 72 hours, included the elderly aircraft-carrier, HMS *Hermes*, which had started life in the late 1950s as a light fleet carrier operating conventional aircraft, been converted to a commando carrier with the removal of her catapults during the late 1970s, and then modified with a 'ski-jump' to operate in the anti-submarine role with British Aerospace Sea Harrier vertical take-off fighters and Westland Sea King helicopters. The other main warship in the task force was the new through deck cruiser or light fleet carrier, HMS *Invincible*. A large number of escort vessels and auxiliaries joined the fleet, under the command of Rear-Admiral Sandy Woodward, while some twenty Sea King helicopters sailed with the initial force for anti-submarine and assault duties.

Throughout the campaign, the helicopter was used for anti-submarine warfare, anti-shipping and plane-guard duties, for assault, transport, communications and liaison, flying-crane and casualty-evacuation duties, as well as flying gunship patrols against Argentinian positions and providing a search and rescue service. The concept of vertical envelopment was hardly used because of the shortage of helicopters and the need to maintain the advantage of surprise in attacks on enemy positions, but vertical replenishment at sea between ships underway, and the transfer of injured servicemen between shore and hospital ship, or between British and Argentinian hospital ship, was accomplished by helicopter. The strain on the helicopter units was made all the greater by the fact that the nearest friendly base to the Falklands was the American airfield on Ascension Island, 3,500 miles away and beyond the un-refuelled range of RAF Lockheed Hercules transports.

The flexibility of the helicopter, with its vertical take-off ability, enabled the British forces to overcome the shortage of suitable vessels to supply the task force and to transport the large numbers of men and equipment needed for the retaking of the islands. Cuts in the British defence budgets had denuded the Royal Navy, which during the mid-1960s had had five conventional aircraft-carriers, including two large carriers, HMS *Eagle* and *Ark Royal*, and two commando carriers, HMS *Albion* and *Bulwark*, although the assault ships HMS *Fearless* and *Intrepid*, scheduled to be withdrawn under more recent defence cuts

On their way to the
Falklands. Royal
Marine Commandos
practising rapid roping
drill from a Westland
Sea King HC4 onto
the helicopter
platform of the
converted cruise ship,
Canberra.

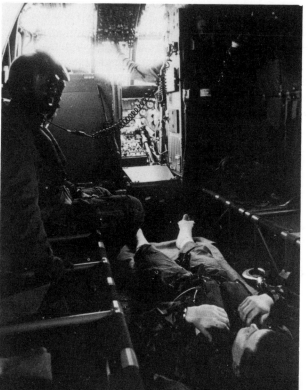

CASEVAC, an inside
shot from the tail-end
of a Sea King HC4.

were still available to the task force. To be fair, even without the cuts, some requisitioning of merchant vessels would have been necessary, but the numbers required were inflated by the cuts, and while dockyard personnel worked wonders to convert vessels, adding helicopter landing-platforms to ships such as the liners *Queen Elizabeth II*, *Canberra* and *Uganda*, and modifying a container ship, the *Atlantic Conveyor*, to carry helicopters and Harrier jet fighters, obviously the anti-aircraft radar and gunnery defences of purpose-built warships could not be fitted in the time available. Sadly, the *Atlantic Conveyor*, one of the unusual Atlantic Container Line container ships which also include a roll-on/roll-off capability with a stern-loading ramp, was attacked and eventually sank after being hit by an air-launched Exocet guided missile fired from an Argentinian Navy shore-based Dassault Super Etendard fighter. Although the Sea Harriers and RAF Harriers carried aboard the ship to help replenish British losses during the opening period of the campaign were all safely aboard the aircraft-carriers at the time of the attack, all but one of the large heavy-lift RAF Boeing-Vertol Chinook helicopters and many of the smaller Westland-Aerospatiale Pumas were lost in the attack, although some at least of these machines were safely ashore before the attack.

A small number of vessels with the advance guard of the task force was dispatched from the rest of the fleet to regain South Georgia. On 21st April, a small force of men from the British Army's crack special services unit, the 22nd Special Air Services Regiment, were landed on the Fortuna Glacier to reconnoitre Argentine positions, but in severe cold and 100-mile-an-hour winds, their tents and equipment were blown away, and the following day they were compelled to ask for a move to a more practical position. A Westland Wessex helicopter flew through appalling 'white out' conditions, in which the pilot could see little, and picked up the men, before crashing on take-off. A second helicopter flew out to pick up the SAS team and the helicopter crew, but this machine, also a Wessex 5, crashed on take-off. Yet a third Wessex, flown by Lieutenant-Commander Ian Stanley from the guided missile destroyer, HMS *Antrim*, then flew to South Georgia and taking off with a massive overload of seventeen soldiers and naval airmen, flew them safely back to his ship: for this achievement, he was awarded the DSO, Distinguished Service Order.

Undaunted, the SAS men attempted a landing later that day using five small Gemini inflatable craft, disliked for their unreliable outboard motors. Three of these reached shore safely, one broke down and the crew was rescued by a helicopter and the fifth broke down and

One Chinook survived the sinking of the *Atlantic Conveyor*, it is seen here flying over a Falklands settlement, possibly Goose Green.

Elderly Westland Scout helicopters of the British Army Air Corps provided sterling service in the Falklands, here is a Scout with a paratrooper ready to jump, while on the other side the CASEVAC stretcher can be seen.

drifted away. The small craft remained missing, with its crew, for several days, before eventually a helicopter homed onto a distress beacon on 26th April, four days later, to find the SAS men on the southernmost tip of South Georgia, having paddled ashore at this, the last land between the island and the Antarctic! Meanwhile, other SAS men and a team from their Royal Marine equivalent, the Special Boat Squadron, had also been landed by helicopter, and eventually a small force, totalling 120 men, retook the principal settlement on South Georgia, Gritvyken, without a shot being fired, taking prisoner an Argentine force of almost 200 men. Later, the British forces retook other Argentinian units on the island.

The eventful, if peaceful, recovery of South Georgia by ground forces, was in contrast to events in the air and on the sea, when a Westland Lynx helicopter on a reconnaissance mission discovered the Argentine Navy submarine, *Sante Fe*, on the surface offshore, and dropped a depth-charge. Later, two Lynx helicopters from the Type 22 frigate, HMS *Brilliant*, attacked the *Sante Fe* with anti-submarine weapons, including a torpedo, causing serious damage, following this with a rocket attack, so that eventually the crippled submarine beached in Gritvyken Harbour.

There were other battles before the recovery of the Falklands. These included commando raids on Argentine positions, of which the most notable was that on Pebble Island, off the north coast of West Falkland, on the night of 14th and 15th May, in which many Argentinian aircraft, including the Pucara ground-attack aircraft, were destroyed along with a radar station and supplies of fuel and ammunition. However, the main attack came a few days later, at Port San Carlos, where the British forces went ashore in large numbers on 19th May, establishing a firm beach-head on a neglected part of the Falkland Sound. The main landing was by assault craft, but afterwards, helicopters were used to resupply the troops ashore and to provide an anti-submarine defence for the landing fleet.

In spite of the many successes of the helicopters and of the operation generally, such intensive operation in weather conditions which would normally have grounded helicopters, with pilots frequently flying ten hours a day, meant that accidents were inevitable, and indeed, it was accident rather than enemy action which accounted for most of the losses of British helicopters in the air. On passage to the Falklands, the task force lost two Sea King helicopters on anti-submarine patrol, while the most serious accident occurred just before the landing at San Carlos, when eighteen men of the SAS were killed while being transferred by helicopter from the liner *Canberra* to an assault ship, after the helicopter crashed into the sea; another nine men from the SAS and the helicopter's crew were rescued. The Sea King was the most common helicopter in service with the task force, both in its standard AS1 and AS3 versions for ASW duties, and in the Commando variant, although the Royal Navy continued to use the Sea King HC4 classification for this machine rather than the manufacturer's terminology. Other accidents included the loss of a small helicopter, either a Gazelle or Scout, in an accident over the Falklands, killing the four Army personnel on board.

Few British helicopters appear to have been lost to ground fire, but the Argentine forces lost two Pumas and a Bell UH–1 Iroquois to attack by Royal Navy Sea Harrier jet fighters. Naturally, British helicopters were vulnerable to attack by Argentinian Dassault Mirage and Israeli Aircraft Industries Kfir (Dagger) jet fighters as well as both Air Force and Navy Douglas A–4 Skyhawk fighter-bombers, but alert and skilful flying avoided what could so easily have been a massacre as low-flying enemy aircraft raced into the Falklands giving little or no warning to the defenceless helicopters. Whenever possible, helicopters landed in a sheltered spot during an air raid, sometimes ditching their under-slung loads before diving for cover, but others hovered just above the ground, merging into the ground with the slopes of a hill or low mountain behind them, becoming difficult targets for fast-flying fighter aircraft. This use of the terrain was helped by the green camouflage, known as 'jungly', of the Royal Navy and Royal Marines troop-carrying Sea Kings, but even more important, it was evidence of the suitability of the tactics planned for use in a war on NATO's Central Front, with helicopters using the terrain both to avoid attack and to maintain an element of surprise against enemy positions.

Because of the shortage of helicopters compared with the heavy commitments awaiting them, and the need to maintain surprise, British troops advancing from San Carlos across East Falkland towards the capital, Port Stanley, marched, carrying up to 120 lbs of equipment per man. This practice was known as 'yomping' to the Royal Marine Commandos taking the northern route through Teal Inlet, while the 2nd Battalion, the Parachute Regiment, marched south to capture Goose Green, with its heavily defended landing strip, and Darwin. The element of surprise lay in the fact that the Argentinian forces flatly refused to believe that marching with heavy equipment over the boggy terrain of the Falklands in winter could be possible, and this technique enabled just 600 men of '2 Para' to take positions held by almost 2,000 Argentinian troops, in a complete reversal of the normal military doctrine which is that the attacking force needs a three or four to one superiority against the defending forces. Helicopters did play an important part in the taking of Bluff Cove and Fitzroy on the south coast of East Falkland: after a telephone call by a British Army officer ascertained that there were no Argentinian forces there, a flight of helicopter gunships flew into the settlements followed by a small number of troop-carrying helicopters. Thick mist hid this operation from the Argentinian forces, but unfortunately, on 11th June, while troops, supplies and heavy equipment were being landed at Bluff Cove by the landing ships, RFA *Sir Tristram* and *Sir Galahad*, the Argentine Air Force attacked in force, severely damaging *Sir Tristram* and crippling *Sir Galahad* when ammunition on board the latter ship caught fire and blew up. A total of more than fifty men, mainly from the Welsh Guards but including members of both ships' companies, were killed and some seventy injured. In spite of heat, dense smoke and exploding ammunition, helicopters flew to the ships to rescue men from their burning decks and from the water immediately around them. In several instances, helicopters hovered in the midst of dense black smoke, lowering winchmen through the smoke and onto the decks or into the water, while others flew low over the water, using the downdraught from their rotor blades to blow life rafts away from the fires. Survivors were ferried ashore by helicopters, the injured were flown to a field hospital and then on to the hospital ship, SS *Uganda*, a converted educational cruise ship which had been fitted with a helicopter landing-platform during a hasty conversion at Gibraltar. The *Uganda* was later to transfer injured Argentinian prisoners to the *Bahia Paraiso*, by now also a hospital ship, using a helicopter for this operation, which was conducted in a "neutral zone" within the 200-mile defensive zone around the Falklands.

As troops advanced on Port Stanley, ultimately forcing first a cease-fire and then Argentinian surrender at midnight on the night of 14th and 15th June, helicopters carried supplies, ammunition, fuel, in large rubber 'fuel pillows', field guns and Rapier anti-aircraft missile batteries, indeed everything the advancing forces needed, to the high ground overlooking the capital. British troops still tended to have to march, although the 1,600 Argentinian prisoners taken at Goose Green and Darwin were luckier, they had a helicopter ride back to San Carlos, leaving some 250 of their comrades dead after the battle.

Another role for the helicopter became apparent during the conflict. The lack of airborne early warning aircraft to provide over-the-horizon radar cover for the British fleet following the withdrawal of ships capable of operating conventional aircraft, resulted in the need to station warships on radar picket duty well forward of the main fleet. A

One lesson of the Falklands, the need for an airborne early warning helicopter was soon satisfied by converting ASW Sea Kings with the addition of the Thorn-EMI Searchwater radar, completed within weeks of the end of the campaign.

sacrificial role at the best of times, an early casualty was the Type 42 guided-missile destroyer, HMS *Sheffield*, the leadship of her class, which was crippled and then eventually sank in bad weather as a result of the damage received when an air-launched Exocet missile hit her amidships. Later, a sister ship, HMS *Coventry*, and two Type 21 Amazon-class frigates, HMS *Ardent* and *Antelope*, were sunk by Argentinian bombs after their air defences had been overwhelmed by repeated attacks. It became clear, even though some had appreciated the problem some years earlier, that airborne early warning was essential for the fleet, and that operating over such a long distance from shore bases, only the helicopter could provide this role. It remains to be seen whether or not this matter receives the urgent attention which it deserves.

Other eventful moments for the Fleet Air Arm helicopters came when two Lynxes from HMS *Brilliant* fired Sea Skua missiles at Argentinian patrol craft, sinking one and seriously damaging another, and a further craft was severely damaged by another Lynx some time later. One helicopter landed men of the SAS on the mainland of Argentina to sabotage enemy aircraft and to spy on aircraft movements, but crash-landed in Chile on the way back, and was destroyed by its three-man crew, who eventually surrendered to the

Chilean authorities. Chile was officially neutral, although it had a number of difficulties of its own with Argentina. There were rescue missions as well, with one Harrier pilot staying in the sea for some hours to avoid detection by Argentinian forces, and then finally signalling so that he could be rescued by a friendly British helicopter.

There is a postscript to the South Atlantic operation. On 20th June, helicopters were used by the Royal Navy to retake the British island of Thule in the South Sandwich Islands, displacing Argentinian scientists illegally settled there.

By contrast, the Israeli invasion of the Lebanon, an operation intended to drive the Palestinian Liberation Organization away from the area south of Beirut and which brought Israeli forces into contact with Lebanese forces, which started on 6th June and lasted for a week, saw a greater use of gunship, assault and CASEVAC helicopters than in any earlier Arab-Israeli war, but showed little novelty in helicopter application, compared to the war in the South Atlantic. It is clear that the frequent and bloody wars in the Middle East will depend more heavily on the helicopter as time passes, but innovation depends on well-equipped forces applying and countering the latest in technology, or in the case of the airborne early warning helicopter, resurrecting an old idea, which originated, as far as the helicopter is concerned, with the Sikorsky S–56. Such innovation does not tend to occur in the Middle East.

12
TOMORROW'S HELICOPTER?

Although few helicopters have managed to achieve the long number of years in production which characterizes the modern fixed-wing aeroplane, in sharp contrast to the 1920s and 1930s when few aircraft remained in production for more than two or three years, there have been exceptions, a number of helicopters which have managed to do just this. The Bell 47, in production from 1947 until 1974 in the United States, and until 1976 in Italy, must be the prime example, but the Bell 206 JetRanger and the Sikorsky S–61 series have both also remained in production throughout most of the 1960s and 1970s. The helicopter is rapidly approaching the stage at which it too will remain in production for a long period of ten to twenty years as a matter of course, rather than as an exception. This does not mean that the limits of technological development are being reached, although it may mean that each advance is insufficient of an improvement to justify a completely new design, even though it may be incorporated into an existing aircraft type. It is, however, a reflection on the cost of development and on the length of production run required to cover these costs, on the interests of operators wishing to standardize on as few types as possible during the twenty years or so during which an aircraft may remain fully airworthy and who therefore wish to be able to re-order types already in their fleet, and on the fact that new types are inevitably more expensive than earlier models whose development costs have been paid some years back, in an inflationary era.

In short, it follows that if we wish to look at the helicopters of the future, then for the most part we have to look carefully at the helicopters of today, most of which will remain in production throughout the 1980s and many of which may remain in production until the mid-1990s, and all of the current production types should see at least a few of their number survive operationally into the next century. This is no exaggeration, but neither is it defeatism, instead it is a reflection of the rapid development of the helicopter over the relatively short history of the practical helicopter, and it is an indication of the maturity of the concept, which has lost much of its novelty.

The immediate future will see developments of existing machines. As this book is being written, the Sikorsky S–61 series is out of

The latest Super Sea Stallion version of the Sikorsky S–65 will be around for many years yet.

production in the United States, and most overseas production lines, although some are still being built by Westland. However, the manufacturer is under some pressure to resume production of at least the civil models. It may not be worth while for Sikorsky to develop the S–61 further, but the S–65 series remains firmly in production, with a new triple-engined Super Sea Stallion version of the United States Marine Corps which may well create a demand for this helicopter on the main export markets. Bell too is offering updated versions of the JetRanger, including the LongRanger and the military Kiowa, while the Super Puma is continuing to be produced by Aerospatiale as an up-rated and modernized version of the SA.330 Puma. Westland's WG.30 is part of the Lynx family, and may enjoy the level of popularity with military customers which the military Lynx has been denied, and which has firmly created an impression of Westland as a naval helicopter manufacturer, which while not wrong is not entirely true either and is certainly not the way which this manufacturer wants to be seen.

Sikorsky has more than repeated the success of the S–61 with the S–70, UH–60 Black Hawk in US Army service and the SH–60B Sea

Hawk in US Navy service, for this time both services have selected one basic helicopter type. The S–70 was initially developed for the United States Army after winning the US Army's Utility Tactical Transport Aircraft System in 1976, after competing against the Boeing-Vertol YUH–61A prototype during a seven-month 'fly-off' evaluation. In contrast to earlier helicopters, the UH–60 series has a very wide, squat profile, reflecting the requirement that the helicopter should be air portable in transport aircraft as small as the Lockheed C–130 Hercules, which can take a single Black Hawk, while the larger C–141 Starlifter can take two and the giant Lockheed C–5A Galaxy can take no less than six! Another requirement of the design was that it should be able, within reason, to survive on the battlefield, and indeed the UH–60 can cope with direct hits from 7.62-mm armour-piercing bullets while the main rotor blades are supposed to be able to survive a hit from a 23-mm shell.

Given the considerable pressure on the American armed forces, and indeed those of any country and even those within the same alliances, to standardize, obviously the Sikorsky S–70 entered the competition for the United States Navy's Light Airborne Multi-Purpose System, LAMPS, helicopter the following year with an advantage over its rivals. The S–70 won this competition and since 1981 has been entering USN service mainly as a replacement for the Kaman SH–2F Seasprite, although the current frigate and destroyer programme will also have the effect of increasing the number of helicopters in USN service. Known in USN service as the SH–60B Sea Hawk, the naval version differs from the military machine in having automatic rotor blade and tail folding, while equipment includes a magnetic anomaly detector, MAD, and surface search radar, and two Mk. 46 torpedoes can be carried. Surprisingly, neither the military nor the naval version has a retractable undercarriage, in spite of the relatively high speed, by helicopter standards, of 185 mph.

Both helicopters use the twin 1,543-shp General Electric T700–GE–700 turbo-shafts of the pre-production YUH–60 and its derivatives, while the military version can carry up to eleven fully equipped troops as well as three crew members, or, in the flying-crane role, 8,000 lbs of external freight can be carried. There seem to be no immediate plans for a civil version, but such a development might be considered almost inevitable, and future export customers probably include the Royal Australian Navy, and perhaps the Spanish Navy and the Canadian Armed Forces. The range of the S–70 series is some 400 miles.

Rather larger is the proposed new Westland-Agusta EH–101

254

American troops prepare to board their Sikorsky UH–60 Black Hawk on an exercise.

A Sikorsky SH–60 Sea Hawk prepares to land on an American frigate, in less than ideal weather.

An improved helicopter, the proposed Westland-Agusta EH–101 for military, naval and civil use.

helicopter, which will use three General Electric T700–GE–401s, similar to the engines used in the SH–60B but with three instead of two in this helicopter which is being developed as a Sea King replacement, and as a helicopter which will also be able to fill the gap left by the end of civil S–61 production.

EH Industries (EH standing for European Helicopter) is a jointly-owned company formed by the Italian Agusta concern and Westland Aircraft in Britain to design and develop a helicopter for the Royal Navy and the Italian Navy, rather than continue with licensed production of American designs, while also being capable of fulfilling a wide range of military medium-lift requirements and being designed with civilian roles in mind. The triple power unit with three 1,550-shp engines is intended to offer greater reliability, although the new Rolls-Royce Gem power-plant currently under development may be used instead, or offered as an alternative. Intended to be in service during the late 1980s, the helicopter may well find that its market is one which seems to be neglected at the moment by Sikorsky, whose range is either smaller or larger than the EH–101, and by Aerospatiale, while Boeing-Vertol is firmly linked to larger helicopters at the moment. It is early to be precise about the difference between the various versions, but an anti-submarine type could almost certainly have some form of dunking sonar, while the military versions are intended to have a rear cabin ramp, and commercial models seating for up to thirty passengers, with a crew of three or four. Bearing in mind current US production and the relationship between the European manufacturers and the United States, one cannot help but speculate on the possibility of some form of licensed production by Sikorsky or Bell.

Agusta itself is preparing to produce the new A.129 Mangusta, or Mongoose, a light attack helicopter based on the rotor systems of the attractive A.109 executive and utility helicopter, which is itself now available as an anti-tank helicopter. The distinguishing feature of the A.129 is that it has a very angular appearance, which is enhanced by the pilot, who occupies the rear seat behind the observer-gunner, being in a prominently raised position. Two 450-shp Allison 250–C30 turboshafts provide a maximum speed of nearly 200 mph, although it is intended that the Rolls-Royce turbo-shaft will be used on production aircraft. TOW or HOT missiles can be fitted, or unarmed rockets and a 7.62-mm minigun. In common with its American counterparts, the A.129 has stub wings on which armaments can be carried.

While the helicopter gains in carrying ability, there is a worthwhile margin of performance being created which designers are using to build in some measure of protection against direct or indirect hits by anti-aircraft fire, in the form which we have already seen on the S–70 series. Indeed, there have even been tentative plans for armoured battlefield helicopters, but these have been flights of fancy, in that true armour plating would create something approaching an armoured tank, far too heavy for rotary-wing technology to lift! However, greater attention to the ability to survive of the helicopter in the future must be considered an essential, but the weak point is, and will remain, the rotor blades. Because of their small size and the high speed of their rotation, the rotor blades are far more vulnerable than the wings of a fixed-wing aircraft, and the loss of even one rotor blade is nothing short of disaster for the helicopter and those aboard.

The action of the rotor blades and their position has also led some designers to consider some form of ejector seat for helicopters, but this subject is fraught with problems. Downwards ejection is impractical given the low operating height of many helicopters, and sideways ejection impractical because of the forces on the human body and the need to support the body during the rapid acceleration of ejection. Indeed, the only way out is up, and for the helicopter crewman that is barred by the rotor blades. While research continues, it is clear that any form of helicopter ejection will have to be allied to some form of rotor blade separation, which creates problems of design, reliability and of control after the blades have been lost.

257

The other form of safety device needed is a warning system for the proximity of power cables, themselves one of the main causes of accidents to helicopters in the developed nations. It is difficult to see a power cable in poor weather conditions, or even in good visibility against an undulating landscape or the blur of a rotor. Some form of passive system with audible warning of the proximity of a cable might be easily achieved due to the field of the electric current, but something rather more positive, perhaps giving warning of the proximity of rotor blades to an obstacle, is really needed to provide a worthwhile advance in helicopter-operating safety. Some helicopters have been fitted with crude wire-cutting devices, but these are not proof against contact between the rotor blades and wires, or indeed a surge of electricity from a high-voltage cable.

The need to ensure better visibility has also led to the fitting of periscopes to the latest generation of military helicopters, although these are not for improved safety but rather to enable the helicopters to see advancing enemy tank formations while hidden by undulations in the ground, by undergrowth or even forests. Most periscopes are fitted to the top of the rotor spindle, using optic fibre technology to relay an image to the observer of the helicopter, who is also responsible for firing weapons and guiding the pilot towards the target. Some form of television picture is produced, in contrast to the purely optical effect of a submarine periscope. Amongst those helicopters tested with periscopes to date can be included the Bell 206 and Hughes 500 series, and the Westland Lynx.

Noise reduction has also stretched the skill of the designer in recent years. Military helicopters do need noise reduction if the element of surprise is not to be lost, although one doubts if much can be heard above the noise of a moving tank army, and this is particularly important on scouting or reconnaissance duties and on assault operations. Unlike the conventional fixed-wing turbo-jet aircraft, engine noise is but a small part of the total noise level produced by a helicopter, and indeed the rotor blades must account for a major part of the noise of any helicopter design. This is not surprising, for with modern turbo-prop light aircraft, propeller noise often rises to nuisance levels and in the executive aircraft market may well have influenced many purchasing decisions towards executive jets, which are more costly, but more pleasant to travel in. Higher helicopter weights and speeds have increased noise levels, and one major problem must be to restrain rotor blade tip speeds to speeds well below that of sound, an inhibiting influence on the search for higher helicopter maximum speeds. Twin-bladed rotors tend to create a distinctive and slightly uneven sound, which can be accentuated if the rotor is even slightly unbalanced with the effect known as ground resonance. One solution attempted so far has been to increase the number of rotor blades to provide both improved lift and either a lower noise effect or at least a more acceptable noise. There can be little doubt that the noise of the helicopter has prevented it from achieving its full commercial potential, and indeed, the light aircraft market already showing a significant proportion of helicopters for business users, would be still more heavily biased in the helicopter's favour if this could be overcome, while, in spite of rising fuel costs, there would almost certainly be a significant number of inter-city, city centre to city centre, helicopter services in the major industrial and commercial regions, overcoming congestion on surface transport, at the airports and on normal airliner operating levels. An improvement in both noise reduction and fuel consumption as well as other costs may result from experiments to eliminate the tail rotor of smaller helicopters. Trials have been conducted with a specially modified Hughes 500.

Possibly one of the most significant developments and one likely to have a profound influence on future helicopter design is the so-called advancing blade concept, known usually as ABC, which has been pioneered by Sikorsky under a joint programme with the United States Army. The concept is based on a contra-rotating rigid main rotor system, and eliminates the imbalance created by the retreating rotor as rotor and aircraft speeds increase. The advancing blade concept derives its name from the fact that the full load of the helicopter is carried on the advancing edges of the two rotors, with the contra-rotating effect cancelling out the penalties of retreating blade stall,

Improving on the helicopter, the Sikorsky XH–59A advancing blade concept prototype.

since the retreating blades are unloaded. A prototype was built, designated the S–69 by the manufacturer and the XH–59A by the US Army, and first flew in July 1973, although it was involved in an accident later that summer. A second prototype, incorporating a number of changes, flew for the first time on 21st July 1975, and over the next twenty months underwent an extensive programme of flight tests, powered by a single 1,825-shp Pratt and Whitney of Canada PT6T–3 Twin-Pac turbo-shaft, reaching speeds of 184 mph in level flight, 224 mph in a shallow dive, and 46 mph in sideways flight, as well as 35 mph flying backwards, indeed demonstrating that it could do anything a conventional helicopter is supposed to do, and at least as well. The two-seat prototype demonstrated that the system offered operators improved manoeuvrability, low noise, high hover efficiency, simplicity and reliability, with higher speeds at higher altitudes than could be offered by a conventional rotor system. Starting in 1978, the prototype has embarked on a high-speed test programme funded jointly by the United States Army and the United States Navy, with its main power-plant supplemented by two Pratt and Whitney J60–P3A auxiliary turbo-jets of 3,000 lbs total thrust in two fuselage side-mounted pods, and in this form has reached speeds of over 300 mph in sea-level level flight. An interesting advantage of the S–69 is that at high speeds the rotor blade speed is reduced from a normal 650 feet per second, or fps, to just 45 fps above 250 mph, overcoming the problem of ever higher blade tip speeds as aircraft speeds increase, and indeed as the airspeed of the helicopter increases, the need for blade movement is reduced.

Ideally, one could foresee some form of vertical take-off aircraft based on the ABC-type helicopter, with rotor blades stopped at higher speeds, operating as a helicopter for take-off and landing and low-speed flight. Indeed, this may be the only practical means of achieving vertical take-off transport aircraft, overcoming the tendency to have to heavily over-power such machines which so far has made them uneconomic and increasingly so, as fuel prices have soared.

The alternative to the ABC concept is the Bell tilt-rotor Model 301, which has been developed under a contract from the National Aeronautics and Space Administration and the United States Army, to whom it is known as the XV–15. The two crew and nine passenger XV–15 uses two wing tip mounted 1,550-shp Avco Lycoming LTC K–4K turbo-shafts which are tilted to the vertical for take-off, landing and hovering, and to the horizontal for forward flight. The latest in a series of several different Bell vertical take-off aircraft prototypes, the 301 first flew in May 1977, and since that time has established its

No way forward. Rising fuel costs have condemned designs such as the
Dornier Do.31E.

maximum speed at 382 mph, while the maximum range is 512 miles.
Although the research programme is to establish the feasibility of the
concept and the design limitations before production aircraft of larger
size are produced, it is perhaps significant that there have been a
number of tilt-rotor and tilt-wing designs, including a version of the
latter concept by Canadair, and these have not, after some twenty years
of testing, advanced beyond the prototype stage. It is possible that the
transitional stage of flight imposes considerable problems and may also
limit the scope for development into larger aircraft, meaning that the
most likely form of development must be either the advancing blade
concept or some form of compound design. This said, there are always
other environmental and economic pressures which affect aircraft
development, otherwise the Fairey Rotodyne would have entered
production.

The modern helicopter is still harder to fly than the conventional
aeroplane, but automation has reduced the pilots workload
considerably. Modern avionics assist all-weather operation, while anti-
submarine equipment such as the dunking sonar, magnetic anomaly
detectors, sonobuoy systems such as Jezebel and improved surface
radar, all help to provide a potent weapon which can be as effective as a
light frigate costing ten times as much and needing ten times as many
personnel to maintain and operate it. Other helicopters are employed

An XV–15 prepares to take off.

Improving on the helicopter, the Bell XV–15 in flight.

on mine counter-measures activity, although this has never been as universal as the anti-submarine helicopter and appears to be less effective than surface vessels.

Just as the helicopters currently in service or under development are those most likely to be in service for the remainder of the century, the helicopter-borne missile systems which have entered service over the past couple of years are still likely to be in use for many years yet, augmented by a few new systems. The missiles which formed the first generation of anti-tank helicopter armament, such as the Aerospatiale AS.10 and AS.11, are now largely out of service having been replaced by a new generation of missiles, including the Franco-German Euromissile HOT and the American Hughes BGM–71A TOW, which are both tube-launched and optically tracked, normally through the helicopter's periscope, wire-guided missiles with a maximum range of just over two miles. Both are subject to further development to improve performance, with infra-red sighting assistance for operation in poor visibility, and offer some further scope for increase in the size of the warhead to counter recent improvements in the design and armour-plating of Soviet tanks. Both TOW and the Euromissile Milan, which uses as similar firing system to HOT and TOW, are in service with the British Army's helicopter units, and indeed with many other armies world-wide. These missiles are often also available for use from ground vehicles, although some ground-based systems, such as the British Aerospace Swingfire, are not available in helicopter-borne forms.

A third generation of missile is being developed by Britain, France and Germany, ready for service by the end of the decade, but decisions have still to be take on the guidance mechanism, the trajectory and on the warhead, all of which clearly indicates the early stage of development and the dependence on existing systems for several years yet! Indeed, the only definite feature appears to be the decision to develop a missile.

The standard Soviet anti-tank missile on the Mil Mi–24 'Hind' attack helicopters is the AT–2 'Swatter', a wire-guided system, which is being replaced by a new missile over the next few years, the AT–6 'Spiral'. 'Spiral' differs from the NATO missiles and the 'Swatter' by being both fairly large, suggesting a large warhead, and in having a range of more than four miles, but no less significant is the use of a radar-based, semi-active homing system, effectively meaning that the missile rides along the radar beam to its target. This system is most usual for surface-to-air missiles, notably the longer-range variety. It would appear that 'Spiral' presents a significant threat to Western tanks, not least because it can be fired from a considerable distance and because of its size, doubtless intended to counter the Chobham armour of future Western tanks, although no vehicles of this type are yet in service, and nor are they likely to be in service with the British Army for much of the 1980s. The widespread introduction of such a missile, coupled to the overwhelming numerical superiority of the Warsaw Pact forces and the large size of the Soviet tank army in particular, can only be counted as a major gain for the Soviets in both military strength and in the psychological advantage which this bestows. It is clear that the Soviet 'Hind' force would have to be destroyed as a top priority in any conflict.

Anti-submarine helicopters tend to operate with homing torpedoes or depth-charges rather than missiles, although to be fair, torpedoes such as the widely used Mk. 46 are in effect underwater missiles with their own homing devices, and the new generation of lightweight torpedoes being developed by GEC-Marconi (the Stingray) in Britain and also in America are also likely to be helicopter-borne and able to hunt submarines at speeds of up to 70 knots underwater.

The shipboard helicopter is also a frequent platform for anti-shipping missiles. Notable amongst these are the AM.39 air-launched versions of the French Aerospatiale Exocet, a radar-guided missile capable of being launched from a helicopter such as a Super Frelon or a Puma, and with a range of more than thirty miles. A shorter range missile also carried by helicopters and used in limited numbers by Iran and Italy is the Italian Sistel Marte Mariner, capable of being used against small naval vessels over ranges of around six miles. A similar missile is the British Aerospace Sea Skua, although with a range of ten miles or more, and which is carried by the Royal Navy's frigate and destroyer-based Westland Lynx helicopters, replacing the anti-shipping version of the wire-guided AS.12 carried by a few Wasp helicopters. A more potent weapon, in every way a replacement for the AM.39 Exocet rather than just a larger counterpart to the Sea Skua, is the P5T, a possible development of the aircraft-borne BAe Sea Eagle anti-shipping missile, shortly to enter service with the Royal Navy and Royal Air Force, and with a range of more than fifty miles.

Some missiles fired from surface vessels can be subjected to mid-course guidance by helicopters equipped with the right radar. A version of the Oto Melara/Matra Otomat ship-to-ship missile used by the Italian Navy in preference to Exocet, and known as Teseo, can use mid-course guidance from Agusta A.109 or AB.212 helicopters, giving this missile considerable accuracy over its range of 50 miles or so. However, the main user of this type of system is the Soviet Navy,

The Sikorsky S–76 Spirit is also finding its way into many air arms now.

which uses specially equipped Kamov Ka–25 helicopters to provide mid-course guidance for the SS–N–3 'Shaddock' missile, with a range of some 500 miles, and possibly also for a new missile known as the SS–NX–12, which is likely to replace 'Shaddock' during the mid- or late 1980s.

Such an armament makes the helicopter a potent weapon in its own right, but of course when viewed in the way in which it can extend the operational radius of action for the parent ship, and with the ability for the ship and the helicopter to be in different places at once, the role of the helicopter takes on a completely different dimension, for it effectively increases the size of the fleet without the cost and the manpower requirement of additional warships.

While it is plain that the helicopter is important and that considerable effort is being devoted to its development, one should not be misled into expecting too many developments too soon. Indeed, it is a compliment to the modern helicopter that such developments cannot occur immediately, because the existing types are so good, and those which are coming, or the developments of the existing helicopters, will suffice for many years yet. The truth is that we have just had a generation change, with the advent of the Bell 222 and the Hughes AH–69, and such other helicopters as the Sikorsky S–70 and S–76, the Westland WG.30, the Agusta A.109 and, if it enters production, the A.129, and that this change will be completed once the EH–101 enters service. Some time must pass before another generation is due, but each succeeding generation is also likely to remain in service and in production for longer, as capability improves and improvement becomes harder to find, let alone fund.

No doubt, one day there will be ABC helicopters of considerable size, or if not these, then some other form of compound helicopter or even, as an outside chance, a form of tilt-wing machine, but that is likely to be some years away, not until the early 1990s at least, and possibly much later, if ever, since there have been so many good ideas which have failed to be translated into practical production aeroplanes, and we should remember that the helicopter as we know it is an exception to that rule!

INDEX

INDEX

de Havilland, 53
de Havilland Canada, 235
de la Cierva, Juan, 18, 19, 30, 51, 52
de la Landelle, 13
Dieuaide, Emmanuel, 14
Distinguished Service Cross, 245
Djinn, Sud Aviation, 108, 109
Do.31E, Dornier, 261
Doblhoff, 27
Dodds, Major, US Army, 49
dunking sonar, 74, 75, 77, 79, 84, 87,
    98, 139, 176, *181*, 183, 186, 196,
    212, 256, 261

E19/46, Air Ministry Specification,
    52
Eagle, HMS, 88, 95, 136
Egyptian Air Force, 51, 105, 218
Egyptian Army, 183, 218
Egyptian Navy, 218
EH–101, Westland-Agusta, 238,
    254, *256*, 265
Ellehammer, 17
Endurance, HMS, 242
Engadine, RFA, 178
'Enosis', 112
EOKA, 113
Erickson, Cdr Frank, USCG, 33, 34
Etendard/Super Etendard, Dassault,
    245
Exocet, 148, 149, 190, 193, 245, 250,
    264

F–4 Phantom, McDonnell
    Douglas, 167
F–27 Friendship, Fokker, 242
Fa.223 Drache, Focke-Achgelis, 20,
    41, 42, 151
Fa.266 Hormise, Focke-Achgelis, 41
Fa.284, Focke-Achgelis, 42
Fa.330, Focke-Achgelis, 21, *22*
Fairchild, 229
Falklands, 60, 241–4, 246, 248
Farnborough Air Show, 163
Fastnet Yacht Race, 200–2

Fearless, HMS, 100, 101, 243
Fearless class, 100
FH–1100, Fairchild-Hiller, 222, 237
fire brigade action, 92, 100
Fisher, Admiral of the Fleet, Lord,
    RN, 24
Fisons Airwork, 229
Fleet Air Arm, 98, 201
Fleurus, Battle of, 9
Flight International, 169
floods, 80, 81
Floyd Bennett Field, 34, 35, 70, 71
FLOSY (Front for the Liberation of
    Occupied South Yemen), 115
Flyer 1, Wright, 14, 15
flying-crane, 80, 104, 107, 144, 151,
    154, 155, 178, 230, 243
Foch, 84, 199, 210
Focke, Professor Heinrich, 41, 42
Forlanini, 14
Fort Austin, RFA, 178
Fort Dusquesne, RFA, 73
Fort Grange, RFA, 178
Fort Monmouth, 34
Fox, Lt, RN, 202
FRAM (fleet modernization and
    rehabilitation), 194
Franklin aero-engines, 26, 39, 43,
    55, 59
French Air Force, 51, 149, 208
French Army, 58, 104, 109, 136,
    147, 149, 150, 203, 208
French Navy, 188, 193
French Indo-China, 59, 85, 112
Furious, HMS, 24, *25*
Fuerza Aerea Argentina, 221
Fw.61, Focke-Achgelis, 20, 29, 41,
    42

Gannet, Fairey, *75*
General Electric aero-engines, 87,
    98, 104, 156, 157, 160, 162, 183,
    185–7, 254, 256
Ghurkas, Brigade of, 114
Georges Leygues class, 193

278